Nelson Science Year 7 Western Australia Student Book
1st Edition
Rachel Whan
Christopher Huxley
Adam Sloan
Debra Smith
ISBN 9780170472852

Series publisher: Catherine Healy
Publisher: Caroline Williams
Project editor: Alan Stewart
Editor: Catherine Greenwood
Series text design: Leigh Ashforth
Series cover design: Leigh Ashforth
Series designer: Linda Davidson
Cover image: iStock.com/harmonia101
Permissions researcher: Catherine Kerstjens
Production controller: Bradley Smith
Typeset by: MPS Limited

Any URLs contained in this publication were checked for currency during the production process. Note, however, that the publisher cannot vouch for the ongoing currency of URLs.

Acknowledgements
The publisher would like to thank Tintern Grammar for the use of their facilities for some of the images featured in Chapters 1 and 2.

© 2022 Cengage Learning Australia Pty Limited

Copyright Notice
This Work is copyright. No part of this Work may be reproduced, stored in a retrieval system, or transmitted in any form or by any means without prior written permission of the Publisher. Except as permitted under the *Copyright Act 1968,* for example any fair dealing for the purposes of private study, research, criticism or review, subject to certain limitations. These limitations include: Restricting the copying to a maximum of one chapter or 10% of this book, whichever is greater; providing an appropriate notice and warning with the copies of the Work disseminated; taking all reasonable steps to limit access to these copies to people authorised to receive these copies; ensuring you hold the appropriate Licences issued by the
Copyright Agency Limited ("CAL"), supply a remuneration notice to CAL and pay any required fees. For details of CAL licences and remuneration notices please contact CAL at Level 11, 66 Goulburn Street, Sydney NSW 2000,
Tel: (02) 9394 7600, Fax: (02) 9394 7601
Email: info@copyright.com.au
Website: www.copyright.com.au

For product information and technology assistance,
    in Australia call **1300 790 853**;
    in New Zealand call **0800 449 725**

For permission to use material from this text or product, please email
aust.permissions@cengage.com

ISBN 978 0 17 047285 2

Cengage Learning Australia
Level 5, 80 Dorcas Street
Southbank VIC 3006 Australia

Cengage Learning New Zealand
Unit 4B Rosedale Office Park
331 Rosedale Road, Albany, North Shore 0632, NZ

For learning solutions, visit **cengage.com.au**

Printed in Singapore by C.O.S. Printers Pte Ltd.
1 2 3 4 5 6 7 26 25 24 23 22

nelson science.
7

Rachel Whan
Christopher Huxley
Adam Sloan
Debra Smith

LEARNING DISCOVERY
ZOANTHID CORALS

This cover image shows a colony of brightly coloured zoanthids attached to a coral reef. Zoanthids are a type of colonial coral, meaning they form a colony of individual polyps all living together. Coral reefs form some of the world's most productive ecosystems.

Zoanthids belong to the phylum Cnidaria and are part of the animal kingdom. This phylum contains about 11 000 aquatic species, which are mostly found in ocean environments. Animals in the Cnidaria phylum have specialised cells known as cnidocytes, which they use mainly for capturing prey.

WA
Australian Curriculum

FIRST NATIONS AUSTRALIANS GLOSSARY

Country/Place

Spaces mapped out that individuals or groups of First Nations Peoples of Australia occupy and regard as their own and that have varying degrees of spirituality. These spaces include lands, waters and sky.

Cultural narrative

A broad term that encompasses any cultural expression that includes (but is not limited to) knowledge and community values that are central to the identity of a particular group of First Nations Peoples.

Cultural narratives can hold information about almost anything, such as the origins of life, or can teach people about acceptable behaviour and rules, such as caring for Country.

They can take the form of songs, stories, visual arts or performances. 'Cultural narrative' is a more accurate and respectful term than 'myth', 'story' or 'fable'; terms that often diminish their importance.

First Nations Australians

'First' refers to the many nations/cultures who were in Australia before British colonisation. This a collective term that refers to all Aboriginal Peoples and Torres Strait Islander Peoples. The term 'Indigenous Australians' is also used to refer to First Nations Australians.

Nation

A self-governed community of people based on a common language, culture and territory.

Peoples and Nations

We use the plural for these terms because First Nations Australians do not belong to one nation/culture. There are many distinct Peoples and Nations. Also, some Nations consist of distinct clans or groups, so are referred to as Peoples.

ACKNOWLEDGEMENT OF COUNTRY

Nelson acknowledges the Traditional Owners and Custodians of the lands of all First Nations Peoples of Australia. We pay respect to their Elders past and present.

We recognise the continuing connection of First Nations Peoples to the land, air and waters, and thank them for protecting these lands, waters and ecosystems since time immemorial.

Warning – First Nations Australians are advised that this book and associated learning materials may contain images, videos or voices of deceased persons.

Contents

First Nations Australians glossary	iv
Acknowledgement of Country	v
Author and contributors	viii
Nelson Science Learning Ecosystem	ix
How to use this book	x

1 INTRODUCING SCIENCE

Chapter map		2
Big science challenge #1		3
1.1	What is science?	4
1.2	Branches of science	6
1.3	Observations and inferences	8
1.4	Safety in science	12
1.5	Equipment in science	16
1.6	Using a Bunsen burner	20
1.7	Measuring in science	23
1.8	Science as a human endeavour	27
1.9	Science investigations	28
Review		30
Big science challenge project #1		31

2 SCIENCE INVESTIGATIONS

Chapter map		32
Big science challenge #2		33
2.1	Introducing the scientific method	34
2.2	Variables	38
2.3	Question, hypothesis and prediction	41
2.4	Testing a hypothesis	44
2.5	Recording results	48
2.6	Analysing results	50
2.7	Evaluation and conclusion	55
2.8	An example of a scientific report: Growing tomato plants with fertiliser	57
2.9	Science as a human endeavour	62
2.10	Science investigations	63
Review		65
Big science challenge project #2		67

3 MATTER: SOLIDS, LIQUIDS AND GASES

Chapter map		68
Big science challenge #3		69
3.1	Matter	70
3.2	Particle theory of matter	72
3.3	States of matter	74
3.4	Properties of solids	78
3.5	Properties of liquids	80
3.6	Properties of gases	82
3.7	Changing state	84
3.8	Science as a human endeavour	88
3.9	Science investigations	90
Review		93
Big science challenge project #3		95

4 PURE SUBSTANCES AND MIXTURES

Chapter map		96
Big science challenge #4		97
4.1	Classifying matter	98
4.2	Pure substances: going further	102
4.3	Solubility	104
4.4	Solutions	106
4.5	Suspensions	110
4.6	Colloids	112
4.7	Science as a human endeavour	114
4.8	Science investigations	116
Review		118
Big science challenge project #4		121

5 SEPARATING MIXTURES

Chapter map		122
Big science challenge #5		123
5.1	Physical properties	124
5.2	Separating suspensions: sedimentation, centrifuging and decanting	127
5.3	Separating suspensions: filtration	131
5.4	Separating solutions: evaporation, crystallisation and distillation	134
5.5	Other separation techniques: magnetic separation, flocculation and chromatography	139

5.6	First Nations science contexts	142
5.7	Science as a human endeavour	145
5.8	Science investigations	146
	Review	148
	Big science challenge project #5	151

6 CLASSIFYING LIVING THINGS

	Chapter map	152
	Big science challenge #6	153
6.1	Classification	154
6.2	Dichotomous keys	156
6.3	Different species	159
6.4	Linnaean classification of living things	162
6.5	The kingdoms of living things	167
6.6	Naming living things	171
6.7	First Nations science contexts	175
6.8	Science as a human endeavour	177
6.9	Science investigations	179
	Review	183
	Big science challenge project #6	185

7 ECOSYSTEMS

	Chapter map	186
	Big science challenge #7	187
7.1	What is an ecosystem?	188
7.2	Energy in an ecosystem	190
7.3	Reviewing food chains	192
7.4	Food webs	195
7.5	Movement of energy and matter in an ecosystem	200
7.6	Modelling ecosystems with pyramids	202
7.7	First Nations science contexts	206
7.8	Science as a human endeavour	209
7.9	Science investigations	212
	Review	214
	Big science challenge project #7	217

8 FORCES

	Chapter map	218
	Big science challenge #8	219
8.1	What is a force?	220
8.2	Measuring forces	224
8.3	Contact forces	226
8.4	Non-contact forces	228
8.5	Balanced and unbalanced forces	232
8.6	Net force	234
8.7	The effect of mass	236
8.8	Mass and weight	238
8.9	Simple machines	241
8.10	First Nations science contexts	247
8.11	Science as a human endeavour	248
8.12	Science investigations	249
	Review	251
	Big science challenge project #8	253

9 OUR PLACE IN SPACE

	Chapter map	254
	Big science challenge #9	255
9.1	Earth, the Sun and the Moon	256
9.2	Reviewing the rotation of Earth	258
9.3	Earth's revolution around the Sun	260
9.4	Seasons	263
9.5	Phases of the Moon	266
9.6	Eclipses	269
9.7	Tides	273
9.8	First Nations science contexts	276
9.9	Science as a human endeavour	279
9.10	Science investigations	280
	Review	283
	Big science challenge project #9	285

| Glossary | 286 |
| Index | 291 |

Author and contributors

Lead author

Rachel Whan

Contributing authors

Christopher Huxley **Adam Sloan** **Debra Smith**

Consultants

Science communication consultants

Joe Sambono
First Nations curriculum consultant

Dr Silvia Rudmann
Digital learning consultant

Judy Douglas
Literacy consultant

Additional science investigations

Reviewers

Dr Silvia Rudmann, Pete Byrne
Teacher reviewers

Aunty Gail Barrow, Nicole Brown, Christopher Evers, Associate Professor Melitta Hogarth, Carly Jia, Jesse King, Dr Jessa Rogers, Theresa Sainty
Reviewers of First Nations Science Contexts pages

nelson science. Learning Ecosystem

Nelson Science 7 caters to all learners

Nelson Cengage has developed a **Science Learning Progression Framework**, which is the foundation for Nelson's Science 7–10 series. An editable version is available on Nelson MindTap.

Reinforce
Nelson MindTap provides a wealth of differentiated activities and resources to meet the needs of all students

Evaluate prior knowledge
Students complete a quiz to test their prior knowledge

nelson science.
Nelson MindTap

Engage
Each chapter showcases fascinating, real-world science in action, while our hands-on activities, short videos and fun interactives keeps students engaged.

Assess
Allocate and grade assessments using our differentiated end-of-topic tests and summative portfolio assessment tasks in Nelson MindTap.

Practise
Our differentiated, scaffolded activities and investigations allow all learners to build essential skills and knowledge

LEARN · LEARN · LEARN · LEARN · LEARN

Nelson MindTap

A flexible and easy-to-use online learning space that provides students with engaging, tailored learning experiences.

- Includes an eText with integrated interactives and online assessment.
- Margin links in the student book signpost multimedia student resources found on MindTap.

Video activity
Cells

For students:
- Short, engaging videos with fun quizzes that bring science to life.
- Interactive activities, simulations and animations that help you develop your science skills and knowledge.
- Content, feedback and support that you can access as you need it, which allows you to take control of your own learning.

For teachers:
- 100% modular, flexible courses let you adapt the content to your students' needs.
- Differentiated activities and assessments can be assigned directly to the student, or the whole class.
- You can monitor progress using assessment tools like Gradebook and Reports.
- Integrate content and assessment directly within your school's LMS.

How to use this book

Big science, real context: The opening page begins the chapter by placing the science topic into a real-life context that is both interesting and relevant to students' lives.

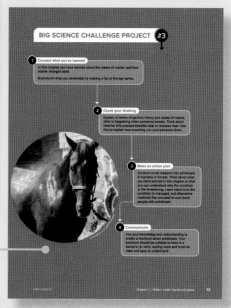

Think, do, communicate: You are encouraged to reflect on and apply your learning to a set of activities, which allows you to make meaningful connections with the content and skills you have just learned.

Learning modules: Content is chunked into key concepts for effective teaching and learning.

Learning objectives: Clear, concise learning objectives give you oversight of what you are learning and set you up for success.

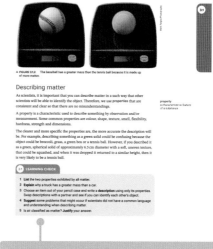

Key words: Key words are defined the first time they appear.

Learning check: These are engaging activities to check your understanding. Activities are presented in order of increasing complexity to help you confidently achieve the module's learning objectives. **Bolded** cognitive verbs help you clearly identify what is required of you. Activities are presented in order of increasing complexity.

Science as a Human Endeavour: Elaborations are explicitly addressed with interesting, contemporary content and activities.

First Nations Science Contexts: This content was developed in consultation with a First Nations Australian curriculum specialist. It showcases the key Aboriginal and Torres Strait Islander History and Cultural Elaborations, with authentic, engaging and culturally appropriate science content.

Science skills in focus: Each chapter focuses on a specific science investigation skill. This is explained and modelled with our Science skills in a minute animation, before you put it into practice in a science investigation. The science skill is reinforced with our Science skills in practice digital activities.

Activities: Activities are open-ended and practical, allowing you to understand the First Nations cultural and historical connections to science.

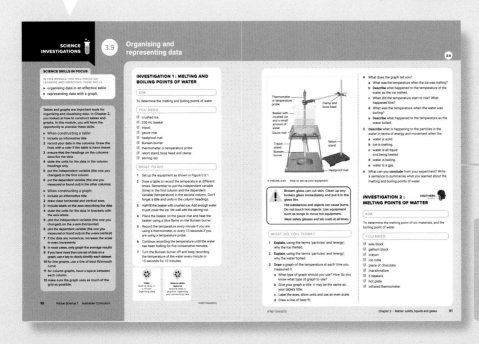

Investigations: Practise and reinforce good scientific method through fit-for-purpose, hands-on science investigations.

How to use this book xi

1 Introducing science

1.1 What is science? (p. 4)
Science uses knowledge to help us understand the world.

1.2 Branches of science (p. 6)
There are many different branches of science, including biology, chemistry and physics.

1.3 Observations and inferences (p. 8)
Observations, both qualitative and quantitative, allow us to make inferences.

1.4 Safety in science (p. 12)
Specific rules and equipment help us stay safe in the science laboratory.

1.5 Equipment in science (p. 16)
Scientists use specialised equipment to conduct experiments.

1.6 Using a Bunsen burner (p. 20)
It is important to know how to use a Bunsen burner safely.

1.7 Measuring in science (p. 23)
Using the correct piece of equipment allows us to make accurate measurements.

1.8 SCIENCE AS A HUMAN ENDEAVOUR: From burrs to Velcro (p. 27)
Observations and questions can lead to scientific advances.

1.9 SCIENCE INVESTIGATIONS: Practise using laboratory equipment (p. 28)
Measurements when ice is added to water

BIG SCIENCE CHALLENGE #1

▲ FIGURE 1.0.1 Elizabeth Blackburn is an Australian scientist who received the Nobel Prize in Physiology or Medicine in 2009.

The Nobel Prizes are a series of five prizes in physics, chemistry, physiology or medicine, literature and peace that were established after the death of Alfred Nobel. In his will, Alfred Nobel left the largest share of his fortune for 'prizes to those who, during the preceding year, have conferred the greatest benefit to humankind'.

In establishing the prizes, Alfred Nobel was recognising that our discoveries in science, and other areas, can have a positive impact on the future. In this chapter, you will learn what science is and some of the methods and equipment used in science to make advances in our understanding.

▶ Who are some recipients of the Nobel Prize that you are aware of?

▶ What are some contributions to humankind that you think are worthy of a Nobel Prize?

#1 SCIENCE CHALLENGE ACCEPTED!

At the end of this chapter, you can complete Big Science Challenge Project #1. You can use the information you learn in this chapter to complete the project.

Assessments
- Prior knowledge quiz
- Chapter review questions
- End-of-chapter test
- Portfolio assessment task: Research project

Videos
- Science skills in a minute: Lab equipment (1.9)
- Video activities: Why science matters (1.1); Australia's award-winning First Nations scientists (1.2); Science safety (1.4)

Science skills resources
- Science skills in practice: Lab vequipment (1.9)

Interactive resources
- Drag and drop: Branches of science (1.2)
- Match: Science equipment (1.5)
- Label: A Bunsen burner (1.6)

Nelson MindTap

To access these resources and many more, visit:
cengage.com.au/nelsonmindtap

Chapter 1 | Introducing science

1.1 What is science?

BY THE END OF THIS MODULE, YOU WILL BE ABLE TO:
- ✓ define science and explain the role of science in our lives
- ✓ give examples of science in your everyday life.

GET THINKING

1. Organise yourself into a group of three or four students.
2. Each student needs something to write on – paper or a new notes page on your device.
3. Set a 2-minute timer (your teacher may help with this).
4. When the timer starts, write as many things as you can think of that have been invented, or discovered, in the last decade.
5. Compare your list with those of other members of your group.
6. Discuss what was needed for each discovery to be possible – was there anything they all had in common?

science
the study of the natural and physical world by asking questions, making predictions, gathering evidence, solving problems and revising knowledge

scientist
a person who uses research to gain knowledge and understanding of any area of science

stereotype
a set idea about something or someone

Science is a system of studying the world

Science is a system of studying the world to solve problems. More specifically, it is a system that uses observation and experimentation to gain knowledge and understanding of the world.

Scientists study many different things and use a wide variety of methods. When we think of scientists, we often picture a **stereotype** of what scientists look like and where they work – mad professors in white coats, working in a laboratory full of glass beakers, jars of chemicals and microscopes. However, scientists work in many different locations, and they wear clothes and equipment that suits a particular purpose. For example, some scientists work in factories, on farms or in hospitals. Others work in or near water, around rocks or ice, at a zoo or even in outer space. Some scientists even spend their working life in an office, using a computer to analyse data or to create models.

▲ **FIGURE 1.1.1** Scientists work in a range of different places: (a) under water, (b) in a zoo, (c) in a mine, (d) in a laboratory.

Science is an ongoing study, with scientists always learning more. Sometimes, discoveries are about new things, but sometimes they show us that our previous understanding wasn't quite right. In these cases, scientists adjust and update models, theories and laws to reflect the new information. This isn't something new, though. People used to think that Earth was flat because their observations were limited to how it looked to them. However, more than 2500 years ago, scientists noticed that some things didn't make sense if Earth was flat. For example, why didn't travellers fall off the end of Earth? Or why did the bottom of a ship disappear first when it sailed away? Then Socrates, a Greek philosopher, saw that the shadow during a lunar eclipse was curved and he proposed that this was because Earth was round. Over time, more and more scientists made observations and measurements that showed Earth is round until it became widely accepted.

More recently, scientists have applied new knowledge to develop vaccines for the virus that causes COVID-19. Traditional vaccines had not been effective against the virus. Scientists worked together internationally to share information. They created and tested new types of vaccines that have helped millions of people to fight the COVID-19 pandemic.

▲ **FIGURE 1.1.2** (a) The curved shadow from Earth during a lunar eclipse was evidence supporting that (b) Earth is round.

1.1 LEARNING CHECK

1 **Define** 'science' and 'scientist'.
2 When you were very young, you learned to write the letters of the alphabet. Then you learned to combine letters to make words, before using words to write sentences. **Explain** how this is similar to scientists updating their understanding as they gain new knowledge.
3 Do you think scientists will ever stop making new discoveries? Give a reason for your answer.
4 Use the Internet to explore some recent advances made by science. The weblinks on this page are good starting points. Focus on one idea and **create** an advertisement in the form of a poster to 'sell' the idea to the rest of the class.

Weblinks
Science daily
New Scientist
Scientific American
ABC Science
CSIRO

Video activity
Why science matters

1.2 Branches of science

BY THE END OF THIS MODULE, YOU WILL BE ABLE TO:
- ✓ describe different branches of science
- ✓ define astronomy, biology, chemistry, environmental science, geology and physics
- ✓ explain how scientists work.

Video activity
Australia's award-winning First Nations scientists

Interactive resource
Drag and drop: Branches of science

GET THINKING

Scientists everywhere use common terminologies when describing the different areas of science so that everyone has the same understanding. Therefore, an important part of learning science is learning the meaning of key terms. One way to learn the definitions of terms is to use flashcards.

1. Choose how you will make your flashcards; for example, on paper or card, or by using an app such as Quizlet.
2. Make a set of flashcards for any branches of science that you already know.
3. As you work through this module, add to your flashcards until you have a set that includes all the branches of science that your teacher wants you to learn.
4. Use your flashcards to practise remembering the definitions for the branches of science.

Scientists at work

Scientists work in all areas of the world around us. There are so many different things to study that the different areas, or branches, have specific names. Some key branches of science are:

biology
the study of living things

chemistry
the study of the composition and properties of matter

geology
the study of the liquid and solid parts of Earth

physics
the study of matter, energy and the interaction between them

astronomy
the study of objects beyond Earth, including stars, other planets and galaxies

environmental science
the study of the conditions of the environment and their effects on all organisms

- **biology** – the study of living things
- **chemistry** – the study of the composition and properties of matter
- **geology** – the study of the liquid and solid parts of Earth
- **physics** – the study of matter, energy and the interaction between them
- **astronomy** – the study of objects beyond Earth, including stars, other planets and galaxies.

Working across different branches of science

Scientists often work in areas that involve more than one branch of science. Sometimes these areas are given their own name, which reflects the branches of science that they use. For example, biochemistry is the study of the chemicals and their processes in living organisms. Other examples are:

- **environmental science** – the study of the conditions of the environment and their effects on all organisms
- geochemistry – the study of the chemical composition of Earth
- astrophysics – the study of how the stars, planets and other celestial objects work
- marine biology – the study of organisms in saltwater environments.

▲ FIGURE 1.2.1   (a) Biology, (b) chemistry and (c) physics are branches of science.

Scientists are named according to the area of science that they study, so that it is clear what they do. For example, a chemist studies chemistry and a biologist studies biology. Some scientists study a very specific area of science and so their name reflects this. For example, a neurologist studies the nervous system and a vulcanologist studies volcanoes.

1.2 LEARNING CHECK

1. What branch of science studies living things?
2. What branch of science will you be studying when you learn about energy?
3. What would a scientist who studies astronomy be called?
4. Over the last few decades, scientists have worked together to understand climate change. What areas of science have contributed to our current knowledge? For each area, **suggest** what scientists would have learned or contributed.
5. Scientists work in many different fields of science. The following list represents just a few of these scientists.

    - Microbiologist
    - Ecologist
    - Anthropologist
    - Entomologist
    - Palaeontologist
    - Haematologist
    - Lepidopterist
    - Oceanologist
    - Agronomist
    - Apiologist
    - Archaeologist
    - Zoologist
    - Botanist
    - Marine biologist
    - Pharmacologist
    - Geneticist

    a. Find out what each of the scientists do. Record your answers in a table.
    b. Did you notice that most of the names in the list end in '-ologist'? Find out what this means. Find two other '-ologist' scientists to add to your list.
    c. Choose one of the types of scientists from your list to conduct further research.
        i. What does this type of scientist study?
        ii. If you became this type of scientist, where might you work and what might you do in a typical day?
        iii. Who is a famous scientist working in this field? What did they discover?
        iv. **Present** your findings in an infographic.

1.3 Observations and inferences

BY THE END OF THIS MODULE, YOU WILL BE ABLE TO:
- ✓ define observation, inference, qualitative data and quantitative data
- ✓ list the senses that are used to collect observations
- ✓ give examples of observations and inferences in different scenarios
- ✓ classify statements as either observations or inferences
- ✓ classify data as qualitative or quantitative
- ✓ use observations to make inferences.

Quiz
Observations and inferences

GET THINKING

How well can you describe your pencil case?

Imagine that you lost your pencil case and needed to describe it to your teacher so that it can be identified from all the pencil cases in lost property. Write a description that is thorough enough for there to be no doubt which pencil case is yours.

Observations

observation
data collected through the senses (sight, smell, taste, touch or hearing) or with measuring tools

An **observation** is any information gained directly from our senses or measuring instruments. You will learn about measuring instruments in Module 1.7.

In primary school you learned about the senses – sight, smell, taste, touch and hearing. We use our senses in science to collect information about whatever we are studying. For example, a scientist who is studying chemistry would make observations such as the colour of a substance, whether bubbles are produced, the odour of a liquid and whether the container gets hotter or colder.

Qualitative data

qualitative data
non-numerical information that relates to a quality, type, choice or opinion

When we use our senses to collect data, we are making a qualitative observation. The information that we collect is called **qualitative data** and is usually in the form of a description. The following are all examples of qualitative data.

- A dark blue colour
- A sweet smell
- Feels hot to touch
- A colourless liquid
- A smooth surface
- Able to be squashed
- A loud bang

Quantitative data

quantitative data
numerical information that is counted or measured and expressed as numbers

Quantitative data is a number. This may be a counting number, such as five petals, or a measurement, such as 14 cm long. Other examples of quantitative data are the temperature of a pond, the time it takes for rubbish to decompose, the volume of oxygen produced by plants and the force needed to bend a piece of metal.

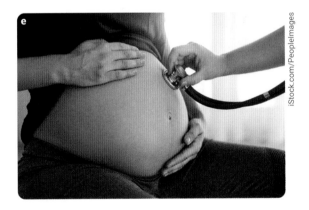

▲ **FIGURE 1.3.1** Scientists use their senses to make observations: (**a**) seeing the shape of grass seeds; (**b**) conducting a blind smell test; (**c**) tasting different brands of food; (**d**) touching tomatoes; (**e**) listening to the heartbeat of an unborn baby.

▲ **FIGURE 1.3.2** A scientist collecting data such as temperature and moisture content in a sustainable ecosystem research project

▼ **TABLE 1.3.1** The differences between qualitative and quantitative data

	Qualitative data	**Quantitative data**
Description	• Think: *qualit*ative = *quality* • This data *describes* what is seen and observed. It does not measure anything. It looks at things such as colours, textures, smells, tastes, appearances and perceptions.	• Think: *quantit*ative = *quantity* • This data deals with numbers and things that can be *counted* or *measured* and represented in tables and graphs.
Example 1: A café	• The walls have a rough texture. • The cabinets have glass doors and stainless-steel counters. • There are wooden tables and chairs with leather seats. • There are floral curtains on the windows.	• The café measures 10 m × 15 m. • There are 4 separate glass cabinets, each measuring 2 m × 1 m × 1 m. • There are 12 circular tables with a diameter of 75 cm. • The front window measures 150 cm × 180 cm.
Example 2: A Year 7 Science class	• The students are friendly. • There is a mixture of boys and girls. • The students are noisy and playful. • The students wear a blue uniform.	• There are 26 students in the class. • There are 14 girls and 12 boys. • The students are aged 12, 13 or 14.
Example 3: A birthday party	• There are both boys and girls. • The party is being held at a park on a mild day. • There is a chocolate birthday cake, party pies, sausage rolls and cupcakes. • There are balloons, loud music and streamers. • There is cordial or water to drink.	• There are 8 boys and 6 girls. • The temperature outdoors is 23°C. • There is 1 birthday cake with 8 candles. There are 24 sausage rolls, 36 party pies and 24 cupcakes. • There are 40 balloons. • There are 6 L of red cordial, 2 L of green cordial and 6 L of water.

Inferences

When you hear a siren, you might guess that there is a fire nearby. You have used an observation (hearing the siren) to make a conclusion about what has happened (a fire). A conclusion based on observations is called an **inference**. If we use more observations to make an inference, it is more likely that the inference will be correct. For example, although hearing a siren might mean a fire engine is going to a fire, it is also possible that it is a police vehicle going to a car accident or an ambulance rushing someone to hospital. However, if it was a very hot day and you could smell smoke, then it is more likely that an inference relating to a fire is correct. This is why it is important that scientists collect as much data as possible before making an inference.

inference
a reasonable conclusion based on observations

1.3 LEARNING CHECK

1 List five observations about a freshly baked chocolate cake.
2 List four observations and one inference from Figure 1.3.3.

▲ FIGURE 1.3.3   What can you infer from this photo?

3 On Christmas Day 2021, the temperature in Perth, Western Australia, reached 42.9°C. Is this quantitative or qualitative data? Use the relevant definition to support your answer.
4 During a police investigation of a burglary, a detective noted that a window frame was damaged. Write an inference that could be made from this observation.
5 **Explain** how observations and inferences are connected.
6 **Discuss** why it is important that doctors make accurate observations when examining a patient.
7 Go to Nelson MindTap and open the supplementary information for this question. Click through the sequence of images and observe what is underneath for each of the areas 1, 2, 3 and 4. What inferences can you make for each area? **Explain** why you needed to observe all four areas before your inference could be correct.

Supplementary information
Learning check 1.3: Question 7, areas 1-4

1.4 Safety in science

BY THE END OF THIS MODULE, YOU WILL BE ABLE TO:
- ✓ state common safety rules for a science laboratory
- ✓ classify actions as either safe or unsafe for a science laboratory
- ✓ explain the purpose of each safety rule in a science laboratory
- ✓ justify why certain actions are unsafe in a science laboratory
- ✓ list common safety equipment for a science laboratory.

Video activity
Science safety

GET THINKING

Safety is always the highest priority in our homes, schools, sporting groups, roads, shops and playgrounds, as well as in science laboratories. Choose one aspect of your everyday life and reflect on the rules that are in place to help keep you safe.

1. List five rules that help keep you safe.
2. Are there any signs that remind us of the safety rules? If there are, draw one and describe how it helps keep people safe.
3. Suggest what would happen if there were no safety rules.

Hazards in the laboratory

In a science laboratory, there are objects and activities that could be dangerous. These are called **hazards**. For this reason, there are specific safety rules and equipment in a science laboratory.

hazard
something that has the potential to harm

Safety rules

Each science laboratory will have its own set of rules. However, there are some rules that are common to all science laboratories.

▲ **FIGURE 1.4.1** Wear safety glasses, lab coat and gloves when doing experiments in a science laboratory.

1. *Follow your teacher's instructions and classroom rules.*

   This is so that you do things safely and in a scientific way.

2. *Do not enter the laboratory without your teacher.*

   Your teacher can give you instructions to ensure that you are not exposed to unnecessary risks in the science laboratory.

3. *Wear safety glass, lab coat (or apron), gloves and enclosed shoes.*

   This protects your eyes and skin from chemicals that could cause damage.

4. *Tie long hair back and secure loose clothing.*

   This stops your hair and clothing catching fire or falling into chemicals.

5. *Do not eat or drink in the laboratory.*

   Food and liquids can be contaminated in the laboratory and therefore should never be eaten in the laboratory.

6. *Do not run in the laboratory.*

   When you run, you are more likely to trip and fall, or knock someone or something.

7. *Do not smell or taste anything in a laboratory.*

   The chemicals in a laboratory can be dangerous to smell or ingest.

8. *Ensure equipment cannot fall off the bench.*

   It is easy for science equipment to fall or be knocked off the bench. This can cause the contents of the container to spill and lead to broken glass, which can cut people.

9. *Report spills and breaks.*

   You teacher needs to know about any spills or breaks. They will check the spill or breakage and direct you so that it is cleaned up appropriately. Any broken glass should be put into a special glass bin.

10. *Never leave a lit Bunsen burner unattended.*

    A lit Bunsen burner can cause fires, or a gas leak if the flame goes out.

11. *Be careful with hot equipment and chemicals.*

    Hot equipment and chemicals can cause burns. To protect against burns, use tongs to hold hot equipment and make sure hot chemicals don't splash onto people.

12. *Always put lids back on bottles.*

    Bottles without lids can easily lead to chemical spills. Keep lids on to reduce this risk.

▲ FIGURE 1.4.2 Ensure that equipment doesn't fall or get knocked over.

▲ FIGURE 1.4.3 Never leave a lit Bunsen burner unattended.

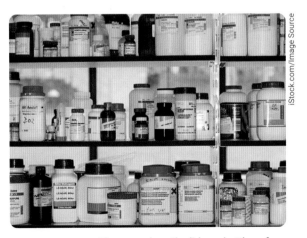

▲ FIGURE 1.4.4 Always replace the lids on bottles of chemicals.

Safety equipment

Science laboratories contain a lot of equipment to keep you safe, including personal protective equipment (PPE) that you will wear. Figures 1.4.5 and 1.4.6 summarise some of the equipment that you will use.

Shower to rinse, dilute and remove chemical spills on people 	Eye wash to rinse chemicals from eyes 	First aid kit to treat small cuts and burns 
Availability of water to dilute and rinse chemicals 	Fire extinguisher to put out any fires 	Fire blanket to smother fires 
Switch to turn off the gas in the case of a gas leak or fire 	Switch to turn off the electricity in the room 	Glass bin for any broken glass 
Spill kit to neutralise, absorb and contain any chemical spills 	Fume hood to remove dangerous gases during experiments 	Tongs to handle hot equipment 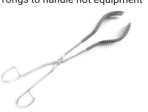

▲ **FIGURE 1.4.5** Some of the safety equipment used in a science laboratory.

▲ FIGURE 1.4.6   Personal protective equipment (PPE) is worn to keep you safe in the laboratory.

1.4 LEARNING CHECK

1. **List** five pieces of equipment that would be used if a beaker containing chemicals was spilled in the laboratory.
2. **Describe** what might happen if someone ate their lunch in the laboratory.
3. **Explain** why it is important to tie long hair back when working in a science laboratory.
4. Organise yourself into groups of three or four students. Use a sheet of paper for the group. Without discussing it, pass the paper around the group, so that each person can write down a safety rule. How many rules can your group remember?
5. Why are the rules in the science laboratory different from those in your maths classroom?
6. **Create** an advertisement to promote one safety rule for the science laboratory. You may do this as a poster, video or radio message.
7. Use three A4 sheets of paper to make 30 cards by cutting each in half lengthwise and then each half into fifths. For each piece of safety equipment in the science laboratory, write its name on one card and draw a diagram of it on another. Turn the cards face down on the table and use them to play a game of Memory.

1.5 Equipment in science

BY THE END OF THIS MODULE, YOU WILL BE ABLE TO:
- ✓ identify common laboratory equipment, including Bunsen burner, beaker, conical flask, test tube, tongs, test-tube rack, retort stand, tripod stand, gauze mat, stirring rod and spatula
- ✓ state the function of each item of common laboratory equipment
- ✓ describe how to use common laboratory equipment
- ✓ compare appropriate laboratory equipment; for example, a beaker and a test tube.

Interactive resource
Match: Science equipment

GET THINKING

What equipment do you already know about? Use the interactive resources on Nelson MindTap to play games to match the names to pictures of different pieces of equipment that are used in the science laboratory. How many did you already know? Did you remember more as you played the games?

Specialised science equipment

equipment
tools used to perform a task

apparatus
equipment designed for a particular use

Scientists use a lot of specialised **equipment**. In most cases, the equipment is designed to perform a particular task and so it is also called the **apparatus**. Science requires accurate, consistent measurements that other scientists can repeat. Therefore, it is important to know the name of the equipment that you have used for your measurements and how to use it safely and accurately. In Module 1.4, you learned about equipment that keeps you safe, and in Module 1.7 you will learn about equipment used to measure data. In this module, you will learn about other equipment used in science.

Equipment used to hold liquids

Different types of containers, including test tubes, beakers and conical flasks (also known as Erlenmeyer flasks), are used to hold liquids depending on their volume and the type of experiment. Test tubes are useful for small volumes, whereas beakers and conical flasks are used for larger volumes. Although beakers and conical flasks have volume measurements on their sides, these are not accurate. Therefore, beakers and conical flasks should not be used as measuring instruments.

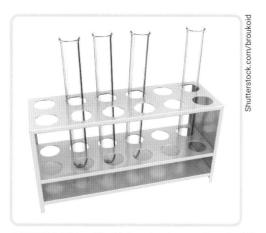

▲ FIGURE 1.5.1 Test tubes, beakers and conical flasks are used for holding liquids.

Equipment used for heating

In some science classes, you will heat substances. You can use either an electric hot plate or a Bunsen burner.

A Bunsen burner sits on a heatproof mat to protect the bench. A tripod stand with gauze mat supports flat-bottomed flasks such as beakers above the flame.

▲ FIGURE 1.5.2 An electric hot plate is used to heat substances.

▲ FIGURE 1.5.3 A Bunsen burner with a heatproof mat, tripod stand and gauze mat

Equipment to help you see small objects

Many things in science are too small to be seen with the naked eye. Magnifying glasses or microscopes **magnify** objects so that you can see them in detail. The apparatus that you choose will depend on how much you need to magnify the object. A microscope will magnify the object much more than a magnifying glass.

magnify
to make something appear larger

▲ FIGURE 1.5.4 Magnifying glasses can let you see small objects, such as parts of a flower, more clearly than with the naked eye.

▲ FIGURE 1.5.5 Light microscopes reveal details of small objects.

Equipment used for holding

There are different-shaped tongs to safely hold items such as test tubes or beakers, especially when they are hot.

Tweezers, or forceps, are used to pick up small things without touching them.

Test-tube racks hold test tubes. They are particularly useful because test tubes have rounded bottoms and so do not stand up on their own. Most test-tube racks can hold multiple test tubes, allowing several tests to be done at the same time.

A retort stand, with a boss head and clamp, holds equipment such as test tubes. It can be used to keep equipment in a particular position, such as above a Bunsen burner.

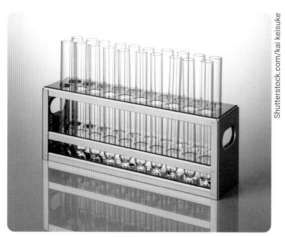

▲ FIGURE 1.5.6 Test-tube racks are used to hold test tubes for experiments.

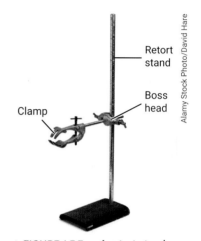

▲ FIGURE 1.5.7 A retort stand, boss head and clamp

Other equipment

Some other pieces of equipment that you will use in the laboratory are a:
- stirring rod – a thin glass or plastic rod used to mix substances
- spatula – a shallow metal or plastic spoon used to transfer powders
- dropper – a small plastic or glass tube used to collect and transfer small volumes of liquid.

▲ FIGURE 1.5.8 A stirring rod, spatula and dropper being used in an experiment

1.5 LEARNING CHECK

1 State the names and uses of the following pieces of equipment.

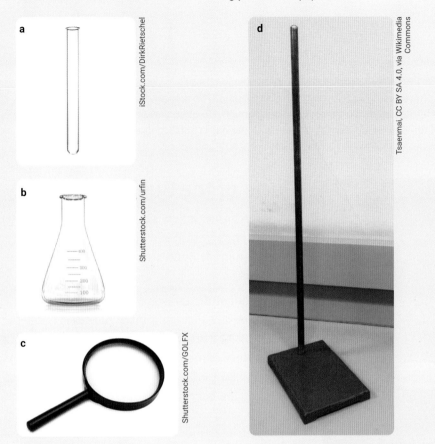

2 **Draw** a labelled diagram to show a beaker, sitting on a gauze mat on a tripod stand, that is being heated by a Bunsen burner sitting on a heatproof mat.

3 **List** two pieces of equipment that could be used to:
   a hold water.
   b pick up something.
   c see something that is very small.
   d heat something.

4 **Describe** the equipment you would use to put some powder in a test tube and heat it safely over a flame.

5 **State** a similarity and a difference between the two items in each of the following pairs:
   a test tube and beaker.
   b microscope and a magnifying glass.
   c spatula and a dropper.
   d retort stand and tongs.

6 **Create** a crossword with clues for the names or uses of science equipment. You may use a website to help create your puzzle.

7 Play a game of Celebrity Heads using the names of science equipment.

1.6 Using a Bunsen burner

BY THE END OF THIS MODULE, YOU WILL BE ABLE TO:
- ✓ state the names and functions of the parts of a Bunsen burner
- ✓ describe how to light a Bunsen burner
- ✓ explain the difference between a blue flame and an orange flame, and when to use each.

Interactive resource
Label: A Bunsen burner

GET THINKING

You can maximise your learning by being prepared. To prepare for learning about the Bunsen burner, take a photo or find a diagram of a Bunsen burner. As you learn about the parts of the Bunsen burner, label them with the name and function. Use a different colour for each part – this will help you remember the information.

Parts of the Bunsen burner

A Bunsen burner is very similar to a gas barbecue: gas burns to produce heat and light. In science, we use the heat from the Bunsen burner to heat objects and chemicals.

The bottom of the Bunsen burner, called the **base**, is flat to ensure that the Bunsen burner is stable on the bench. The gas enters the Bunsen burner through the **gas hose**, which is connected to the **gas tap**. Air enters the Bunsen burner through the **air hole** and travels up the **barrel** with the gas. The oxygen in the air reacts with the gas, producing a flame at the top of the barrel. The amount of air mixing with the gas is controlled by opening or closing the air hole by turning the **collar**. You can see the parts of a Bunsen burner in Figure 1.6.1.

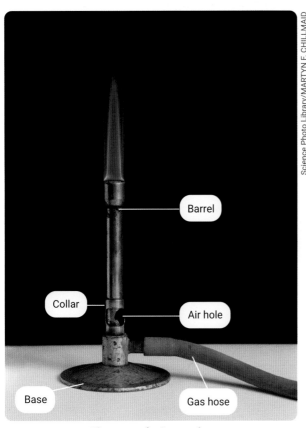

▲ FIGURE 1.6.1   The parts of a Bunsen burner

Types of flames

A Bunsen burner has two types of flames: an orange, or **safety flame**, and a **blue flame**. You can see these two types of flame in Figure 1.6.2.

The orange flame is known as the safety flame because it is easy to see. It is a cooler flame because it occurs when the air hole is closed, meaning that there is less oxygen available. This type of flame produces black soot and will dirty the glassware. Therefore, it is not used when heating substances.

The blue flame is produced when the air hole is open. It is a hotter flame than the orange flame because of the extra oxygen available. Because it is cleaner and hotter, a blue flame is used to heat substances. However, the blue flame is difficult to see, so the flame should always be changed back to the orange safety flame when not being used to heat objects.

safety flame
the cooler flame from a Bunsen burner that is easily visible because it is orange; also known as the orange flame

blue flame
the hottest flame from a Bunsen burner; blue in colour

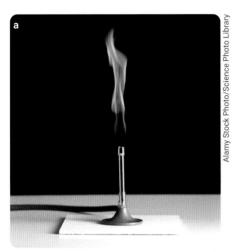 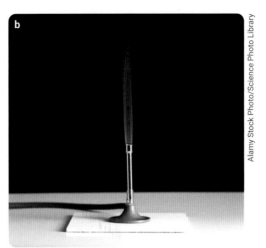

▲ FIGURE 1.6.2 A Bunsen burner with (a) an orange flame (the safety flame) and (b) a blue flame

Lighting a Bunsen burner

The following steps tell you how to safely light a Bunsen burner:

1. Place the Bunsen burner on a heatproof mat.
2. Check that there is no damage to the gas hose.
3. Connect the gas hose securely to the gas tap.
4. Close the air hole.
5. Light the match or gas lighter.
6. Hold the match or gas lighter just above and slightly to the side of the top of the barrel.
7. Turn the gas tap on. An orange flame should appear from the top of the Bunsen burner.
8. When heating objects, turn the collar to open the air hole. This will turn the flame to blue.

Turning off the Bunsen burner

When you have finished using the Bunsen burner, turn the gas tap off. The remaining gas will react and then the flame will go out. This ensures that there isn't any unreacted gas released into the room; this would be dangerous.

Sometimes the flame may get blown out. If this happens, turn the gas tap off immediately to stop gas being released.

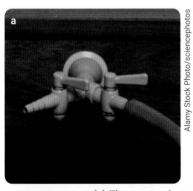 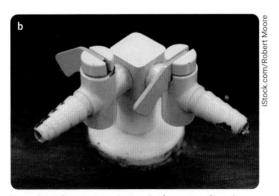

▲ FIGURE 1.6.3 (a) The gas tap is turned on when the tap is pointing in the same direction as the outlet. (b) The gas tap is turned off when the tap is pointed away from (90°) the direction of the outlet

1.6 LEARNING CHECK

1. **Match** each part of the Bunsen burner with its function.

Part	Function
Barrel	Allows the Bunsen burner to sit upright
Gas tap	Carries gas to the Bunsen burner
Base	Controls how much air mixes with the gas
Air hole	Carries the gas up the Bunsen burner
Gas tube	Opens or closes the air hole
Collar	Controls whether gas is released or not

2. Copy and **complete** the table with information about the two types of flames.

Description	Orange flame	Blue flame
Temperature (very hot or hot)		
Clean or dirty (produces soot)		
Safety flame (yes or no)		
Air hole open or closed		
Amount of oxygen (high or low)		

3. **Explain** why it is important that the gas tap is turned on after the match is lit and held above the barrel, and not before.

4. Under your teacher's supervision, set up a Bunsen burner on a heatproof mat with a tripod stand and gauze mat. Put a piece of wax in a beaker and place the beaker on the gauze mat. Light your Bunsen burner and open the air hole to turn it to a blue flame. Heat the wax until it has all melted. Close the air hole to change the flame to an orange flame and then turn the gas off. Leave the equipment to cool then follow your teacher's instructions to pack the equipment away. **Describe** the blue flame and explain why it was used when you were heating the wax.

5. Many people use videos to learn how to do new things. **Create** an instructional video about how to light a Bunsen burner. Your teacher may let you take videos or photos showing the steps or you could use drawings instead. Add voice-overs and text for each step so that it is clear. If you do not have access to a video-recording device, create a simple picture book to show the steps.

1.7 Measuring in science

BY THE END OF THIS MODULE, YOU WILL BE ABLE TO:
- ✓ define parallax error and meniscus
- ✓ identify common laboratory equipment, including thermometer, measuring cylinder, stopwatch and electronic balance
- ✓ label the meniscus on a diagram
- ✓ describe how to reduce parallax error
- ✓ describe how to correctly measure mass, volume, temperature, length and time.

GET THINKING

What do you already know how to measure?
1. How do you measure the amount of flour when baking a cake?
2. How do you measure how much you weigh?
3. How do you measure how tall you are?
4. How do you measure how long it takes you to get to school in the morning?
5. How do you measure how much food to feed your pet?
6. What else do you measure?

Measuring mass

Mass is the measure of the amount of matter in an object. It is measured in grams (g), kilograms (kg) or milligrams (mg). We measure mass with an electronic balance in a process called weighing (Figure 1.7.1).

In science, we usually need to put the object, or chemical, that we are weighing in a container (e.g. a beaker) to hold it. It is important that you don't include the mass of the container in the mass of the object. To avoid this error, the container is put on the balance first and the balance is zeroed or tared. This means that the balance is reset to a mass of zero. Then the object or chemical is added to the container.

mass
the amount of matter in an object, measured in kilograms (kg), grams (g) or milligrams (mg)

▲ FIGURE 1.7.1   The powder weighs 75.648 g.

Measuring the volume of liquids

volume
the amount of space occupied, measured in litres (L) or millilitres (mL)

Volume is described as the amount of space occupied by something. It is measured in litres (L) or millilitres (mL) – there are 1000 millilitres in 1 litre.

A measuring cylinder (also known as a graduated cylinder) is used to accurately measure the volume of a liquid. There are different-sized measuring cylinders. To get the most accurate measurements, use the smallest measuring cylinder that will hold the volume you need to measure.

When using any measuring apparatus, you need to ensure that it is on a flat, level surface. You also need to be at eye level with the measurement that you are reading on equipment such as measuring cylinders and thermometers. This reduces **parallax error**, which is the error that occurs when reading measurements from an angle.

parallax error
an error in the reading of an instrument due to the viewing angle

meniscus
the curved surface of a liquid when it is in a thin tube

When you look at the liquid in a measuring cylinder, you will notice that the surface is curved, not flat. This curved surface is called the **meniscus**. To measure the volume of liquid, read the measurement at the bottom of the meniscus (Figure 1.7.3).

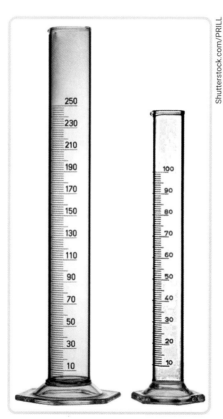

▲ **FIGURE 1.7.2** Measuring cylinders: use the smallest measuring cylinder that will hold the volume you need to measure to get the most accurate measurement.

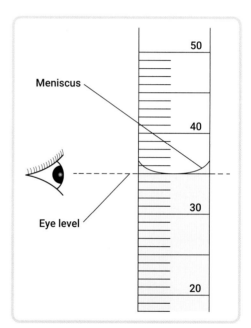

▲ **FIGURE 1.7.3** The meniscus on liquid in a measuring cylinder. The volume of this liquid is 35 mL.

Measuring temperature

Temperature indicates how hot or cold something is. We measure temperature in degrees Celsius (°C) by using a thermometer or temperature probe.

A thermometer consists of a glass tube with a liquid inside and numbers marked on the outside (Figure 1.7.4). When the liquid gets hotter, it expands and moves up the tube. The temperature is the number at the top of the liquid.

A temperature probe is a digital thermometer (Figure 1.7.5). Most temperature probes connect to a device and give a read-out of the temperature.

▲ FIGURE 1.7.4 A thermometer is used to measure the temperature.

▲ FIGURE 1.7.5 A temperature probe measuring the temperature of ice in a beaker

temperature
how hot or cold something is, measured in degrees Celsius (°C)

Measuring length

Length is the distance between two points, measured in metres (m), centimetres (cm) or millimetres (mm). There are 10 mm in 1 cm, 100 cm in 1 m, and 1000 mm in 1 m.

We use a ruler or measuring tape to measure the length of something by placing the zero mark at one end and then reading the number at the other end (Figure 1.7.6). You should look directly at the number to avoid parallax error and to get an accurate measurement.

length
the distance between two points, measured in metres (m), centimetres (cm) or millimetres (mm)

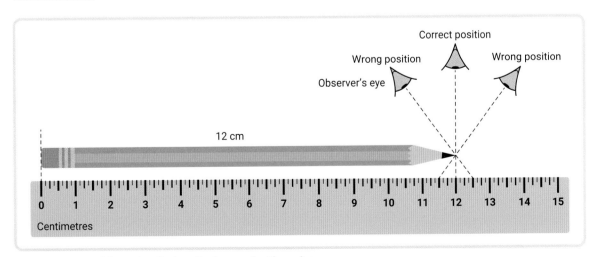

▲ FIGURE 1.7.6 Measuring the length of a pencil with a ruler

Chapter 1 | Introducing science

Measuring time

time
how long something takes, measured in hours (h), minutes (min) and seconds (s)

Time is how long something takes. It is measured in hours (h), minutes (min) and seconds (s). In the past, people used stopwatches to measure time. Now, most electronic devices such as phones and tablets can accurately measure time too.

When measuring time in science, it is important that the timer is started and stopped at the right times.

1.7 LEARNING CHECK

1 Copy and **complete** the table to link the equipment used to measure each quantity.

Quantity	Equipment	Unit
Length		
	Thermometer	
		grams
		millilitres
Time		

2 **Write** a set of instructions for a student to measure the height of a blade of grass.

3 **State** the volume measured in each of the following pieces of equipment.

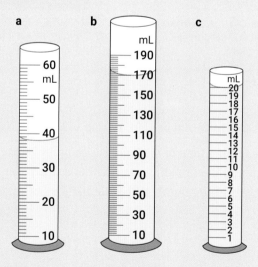

4 **Explain** why you would use a measuring cylinder and not a beaker to accurately measure the volume of a solution.

5 During an experiment, a student turned on an electronic balance and placed a beaker on it. He then used a spatula to add sugar until the balance read 50.0 g. Was there 50.0 g of sugar in the beaker? **Explain** why or why not.

6 Do you think a thermometer or a temperature probe is more accurate? **Justify** your answer.

1.8 From burrs to Velcro

SCIENCE AS A HUMAN ENDEAVOUR

BY THE END OF THIS MODULE, YOU WILL BE ABLE TO:
✓ explain how looking at things from a different perspective can lead to scientific advances.

After a hike in 1941, Swiss electrical engineer George de Mestral came home with burrs stuck to his clothes. Rather than being annoyed, he wondered whether the stickiness of burrs could be adapted to be useful.

De Mestral looked at the burrs under a microscope and saw that they had hooks that latched onto clothes (Figure 1.8.1b). He then spent many years trying to use this structure to make a material with hooks on one side that could latch onto loops on the other side. De Mestral finally succeeded, with Velcro becoming available in the early 1960s. Since then, other companies have produced similar products, which are known as hook and loop fasteners.

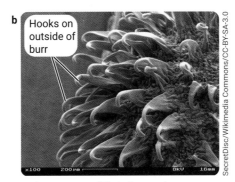

▲ **FIGURE 1.8.1** (a) Burrs stick to clothes and are hard to remove. (b) Under an electron microscope it can be seen that burrs have hooks that cling to fabric.

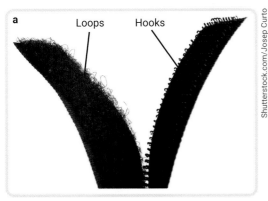
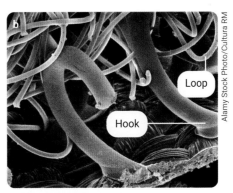

▲ **FIGURE 1.8.2** (a) Velcro is made up of hooks and loops. (b) Velcro under an electron microscope, showing how it has hooks on one side that hold on to loops on the other side.

1.8 LEARNING CHECK

1. Use a dictionary (either online or a book) to find the definition of 'mimicry'. How is the development of Velcro an example of mimicry?
2. What other inventions are an example of mimicry?

Weblinks
Live Science: Velcro
Inventing Velcro
Time: History of Velcro

SCIENCE INVESTIGATIONS

1.9 Practise using laboratory equipment

SCIENCE SKILLS IN FOCUS

IN THIS MODULE, YOU WILL FOCUS ON LEARNING AND IMPROVING THESE SKILLS:

▶ practise using laboratory equipment.

▶ Lab equipment tips

Remember here some general pointers to help you use laboratory equipment correctly and safely.

- Check the materials listed in the You need section. Do you have everything?
- Always ensure equipment is clean and dry before you use it.
- Take care when using fragile equipment, such as glassware.
- Let your teacher know if you accidentally damage or break equipment.
- Make sure you are careful when pouring and measuring liquids.

Video
Science skills in a minute: Lab equipment

Science skills resource
Science skills in practice: Using lab equipment

⚠ Warning

- Broken glass can cut skin. Take care when using glassware. Report any breakages immediately and follow your teacher's instructions to clean it up. Put broken glass in the glass bin.
- Hot substances and objects can cause burns. Do not touch hot objects. Use equipment such as tongs to move hot equipment. Wear safety glasses and lab coats at all times.

MEASUREMENTS WHEN ICE IS ADDED TO WATER

AIM

To use science equipment to measure the temperature change when an ice cube melts in water

YOU NEED

- ☑ water (approximately 150 mL)
- ☑ 2 × 250 mL beakers
- ☑ 200 mL measuring cylinder
- ☑ ice cube made from water with food dye
- ☑ electronic balance
- ☑ thermometer or temperature probe
- ☑ stopwatch or timer on a device

WHAT TO DO

1. Use a measuring cylinder to measure 150 mL of water. Record this volume in the results table (Table 1.9.1).
2. Pour the water into a beaker.
3. Use the thermometer (or temperature probe) to measure the temperature of the water. Record the temperature in Table 1.9.1.
4. Place the second beaker on the electronic balance and press zero (or tare) so that the reading is 0.00 g.
5. Add one ice cube to the beaker to measure its mass. Record the mass in the results table.
6. Prepare the stopwatch (or timer) so that it is ready to start timing.
7. Add the ice cube to the water and start the stopwatch (or timer) (Figure 1.9.1).
8. Stop the stopwatch (or timer) as soon as the ice has melted. Record this time in Table 1.9.1.
9. Use the thermometer (or temperature probe) to measure the temperature of the water when the ice cube has melted. Record this final temperature in Table 1.9.1.
10. Pour the water down the sink and rinse your equipment before packing it away.

◀ **FIGURE 1.9.1** The coloured ice cube is added to the water.

RESULTS

1. Copy and complete the results table (Table 1.9.1).
2. List five observations that you made during the experiment.

▼ **TABLE 1.9.1** Measurements when an ice cube was added to water

Description	Measurement
Volume of water (mL)	
Initial temperature of water (°C)	
Mass of ice cube (g)	
Time for ice cube to melt (min:s)	
Final temperature of water (°C)	

WHAT DO YOU THINK?

1. Why do you think the ice cube melted?
2. How many degrees Celsius did the temperature of the water change by?
3. What do you think would have happened if:
   a you had used a smaller volume of water?
   b the ice cube had a bigger mass?
   c the ice cube had been crushed?
4. What steps did you take so that your measurements were as accurate as possible?

CONCLUSION

What did you learn in this activity?

1 REVIEW

REMEMBERING

1. **State** what is studied in:
   a. biology.
   b. chemistry.
   c. geology.
   d. physics.

2. Give three examples of qualitative data and three examples of quantitative data.

3. **State** five safety rules for working in a science laboratory.

4. Safety glasses, laboratory coats and gloves improve safety while conducting an experiment. **List** two other items in a science laboratory that are used for safety.

5. **Describe** what each of the following are used for.
   a. Electronic balance
   b. Test tube
   c. Gloves
   d. Bunsen burner
   e. Thermometer
   f. Stirring rod
   g. Heatproof mat

6. **Name** parts A–E on the diagram of a Bunsen burner.

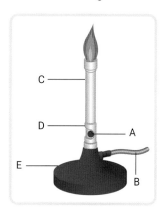

7. **List** two ways that you can stay safe when using a Bunsen burner.

UNDERSTANDING

8. **Describe** when you would use a tripod stand and gauze mat.

9. **Explain** why the blue flame of the Bunsen burner is only used when heating something.

10. **Explain** why it is important to read the scale of science equipment at eye level.

APPLYING

11. **Suggest** what biophysicists study.

12. List three observations and one inference that you can make from the following photo.

EVALUATING

13. **Describe** three personal characteristics of scientists. Explain why each characteristic is important.

14. There are more areas of science now than there were 200 years ago. **Suggest** a reason for this observation.

15. **Justify**, or refute, this statement: 'Winning awards is an incentive for scientists'.

16. John and Judith were conducting the same science experiment. In one step, they needed to measure 50 mL of water. John used a 100 mL beaker and carefully poured water up to the 50 mL line. Judith chose a 25 mL measuring cylinder. She added water up to the 25 mL mark, poured it into a clean beaker and then measured a second 25 mL volume. Whose method was more accurate? **Justify** your answer.

17. In your opinion, which is more reliable: qualitative or quantitative data? **Explain** your reasoning.

CREATING

18. Write a short story about a day in the life of a piece of equipment that is mentioned in this chapter. Think about all the things that it might 'do' during the day, 'who' or what it might interact with and what things might go wrong. You could illustrate your story and, if you have access to a device, even publish it as an ebook.

BIG SCIENCE CHALLENGE PROJECT #1

In 2005, Australians Dr Robin Warren and Professor Barry Marshall were awarded the Nobel Prize in Medicine or Physiology for their discovery about stomach ulcers. Before their discovery, people believed that stomach ulcers were due to lifestyle factors such as diet and stress. Warren and Marshall's research showed a link between the bacteria *Helicobacter pylori* and ulcers, which meant that ulcers could be treated with antibiotics. The discovery had such a positive impact that they were considered worthy of the Nobel Prize.

Warren and Marshall's discovery is one of many examples of scientists making discoveries that have significant benefits to our world. Some of these have been recognised with the Nobel Prize; however, there are many other worthy contenders.

Your challenge is to learn more about one discovery, who made it, how it was made and how it has affected the world. Use the weblinks as starting points for your research.

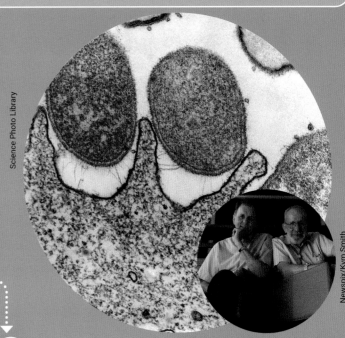

Dr Robin Warren and Professor Barry Marshall (left) and a microscopic view of the stomach (blue) and the bacteria *Helicobacter pylori* (red)

1 Connect what you've learned

In this chapter you have learned about the branches of science, how scientists work, how they stay safe and how they collect their data. Use your research to answer the following questions.

a What is the name of the scientist who made the discovery, when was the discovery made and what country was it made in?

b What branch of science did the scientist work in?

c What was discovered?

d How was the discovery made?

e How has the discovery been helpful?

2 Check your thinking

Think about what made this discovery possible,

- What personal characteristics would the scientist have needed to be successful?
- What challenges might the scientists have needed to overcome?

3 Communicate

Imagine that a new award has been established for the most valuable scientific discovery. Write a letter to the committee that decides the winner to explain why the discovery that you have researched should receive the award.

Weblinks
Nobel Prize winners
Britannica: Nobel Prize
Compound Chemistry: Nobel prizes
Year in review

2 Science investigations

2.1 Introducing the scientific method (p. 34)
The scientific method is a process to obtain valid information in a consistent and repeatable way.

2.2 Variables (p. 38)
Variables are any factors that could change a result. They are classified as independent variable, dependent variable and controlled variables.

2.3 Question, hypothesis and prediction (p. 41)
The hypothesis is related to a question and predicts the relationship between an independent and a dependent variable being tested. A prediction states the expected results.

2.4 Testing a hypothesis (p. 44)
A hypothesis is tested by changing the independent variable to identify its effect on the dependent variable.

2.5 Recording results (p. 48)
Organising results in an effective table allows trends and patterns to be identified.

2.6 Analysing results (p. 50)
Data is analysed for trends and patterns to identify whether the hypothesis is supported. Graphs are tools used to analyse data.

2.7 Evaluation and conclusion (p. 55)
An investigation is evaluated to decide whether the conclusion is valid. A conclusion summarises the relationships found and whether they support the hypothesis.

2.8 An example of a scientific report: Growing tomato plants with fertiliser (p. 57)
Scientists write reports to share the findings of investigations.

2.9 SCIENCE AS A HUMAN ENDEAVOUR: Scientists as communicators (p. 62)
Jane Goodall is a well-known science communicator.

2.10 SCIENCE INVESTIGATIONS: Writing hypotheses (p. 63)
How quickly does bread go mouldy?

BIG SCIENCE CHALLENGE #2

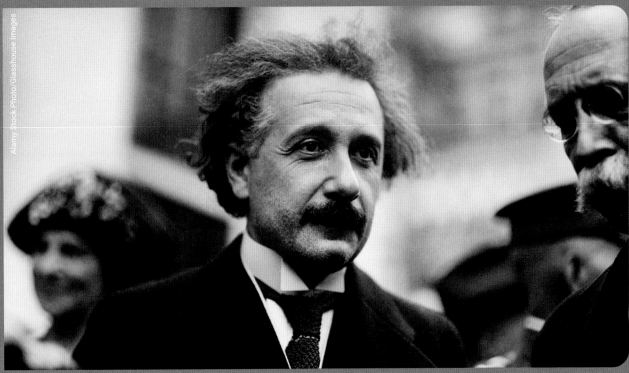

▲ FIGURE 2.0.1 Albert Einstein famously said 'The important thing is to not stop questioning'.

Science is all about answering questions to further our understanding.

▶ What question would you like to answer?

▶ How will you find the answers?

#2 SCIENCE CHALLENGE ACCEPTED!

At the end of this chapter, you can complete Big Science Challenge Project #2. You can use the information you learn in this chapter to complete the project.

Assessments
- Prior knowledge quiz
- Chapter review questions
- End-of-chapter test
- Portfolio assessment task: Science Investigation

Videos
- Science skills in a minute: Hypotheses **(2.10)**
- Video activities: What is the scientific method? **(2.1)**; Fair testing **(2.4)**; Science communication tips **(2.9)**

Science skill resources
- Science skills in practice: Writing hypotheses **(2.10)**
- Extra science investigation: Popcorn hypothesis **(2.3)**

Interactive resources
- Drag and drop: Variables **(2.2)**; Types of data **(2.6)**
- Label: Scientific method steps **(2.1)**; Parts of a scientific report **(2.8)**
- Crossword: The scientific method **(2.7)**
- Match: Data and results **(2.5)**

Nelson MindTap

To access these resources and many more, visit:
cengage.com.au/nelsonmindtap

Chapter 2 | Science investigations

2.1 Introducing the scientific method

BY THE END OF THIS MODULE, YOU WILL BE ABLE TO:
- ✓ list the steps in the scientific method
- ✓ explain the importance of the scientific method.

> **GET THINKING**
>
> Figure 2.1.2 illustrates the steps in the scientific method. Use the first letter of each word to create a saying to remember the steps. The saying does not need to be serious or factual – just something that you will remember. These sayings are called **mnemonics** and are a strategy to help remember a list of words.

mnemonic
a memory aid that uses the pattern of letters in words

The scientific method

In Chapter 1 you learned that scientists are constantly finding answers to questions about the world. When they find these answers, it is important that the results can be trusted.

Imagine if you investigated how quickly the cars in Figure 2.1.1 could accelerate (or increase their speed). Based on the results, you may conclude that red cars are the fastest cars. However, there are several factors, other than colour, that affect how quickly cars accelerate. These factors include the age, shape, mass and size of the car, its motor, and the road surface. Therefore, the investigation was not a **fair test**, and its results cannot be trusted.

fair test
a way of finding the answer to a question that ensures the answer is valid

▲ **FIGURE 2.1.1** A range of cars used to test which colour of car is the fastest

Scientists have devised an approach to conducting scientific investigations and ensuring the test is fair, called the **scientific method**. The scientific method is a series of steps that enables scientists to plan and conduct experiments in a consistent and repeatable way.

scientific method
a systematic way of gaining knowledge

The basic steps of the scientific method are as follows:

1. ask a testable question about an observation
2. formulate a hypothesis (a predicted reason for the observation)
3. test the **hypothesis**
4. collect and analyse the data
5. draw conclusions
6. evaluate the investigation
7. revise the hypothesis and/or repeat the test and/or ask a new question.

hypothesis
a testable explanation for something based on existing knowledge; a testable statement of the predicted relationship between the independent and dependent variables

Part of the scientific method involves reflecting on the results. If the results support the hypothesis, then the test should be repeated to ensure that the same results are achieved. If the results do not support the hypothesis, the hypothesis is revised and new testing can occur. Sometimes an investigation raises new questions that can then be tested.

You will learn more about each part of the scientific method in the following modules.

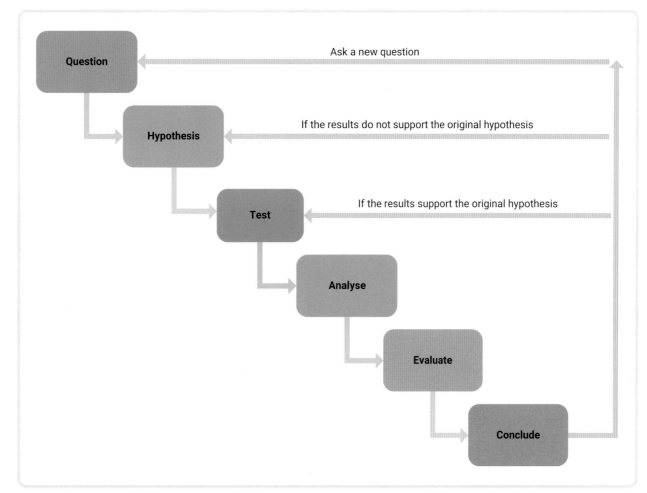

▲ FIGURE 2.1.2   The scientific method

Video activity
What is the scientific method?

Interactive resource
Label: Scientific method steps

Using the scientific method

An example of the scientific method is choosing the shape of a travel coffee mug to keep coffee hot. The steps used are summarised in Figure 2.1.4.

▲ **FIGURE 2.1.3** Which shape of travel coffee mug will keep coffee the hottest?

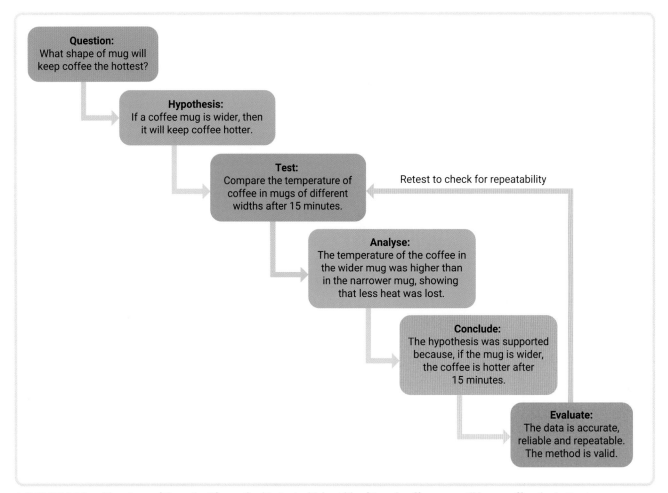

▲ **FIGURE 2.1.4** The steps of the scientific method to test which width of travel coffee mug will keep coffee the hottest

Finding the answers to questions
ACTIVITY 2.1

Outline how you would test each of the following questions so that it is a fair test.

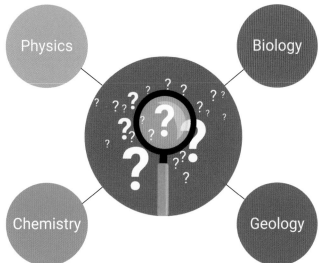

- What colour cars travel the fastest?
- How does the thickness of a string affect the sound it makes on a guitar?
- How does the temperature of honey affect how easy it is to pour?
- What type of soft drink is the most acidic?
- How does the temperature of water affect plant growth?
- How does the intensity of exercise affect heart rate?
- How does the size of particles in a rock affect its strength?
- How does acid rain affect rocks?

2.1 LEARNING CHECK

1. **List** the steps of the scientific method, in order.
2. **Explain** why it is important that a test is conducted more than once.
3. **Outline** the steps that you would take to test which fly repellent is the most effective.
4. The scientific method is used by scientists around the world. **Explain** why it is important for all scientists to use the same process.
5. When a scientist finds new answers, their work is checked by other experts (this is called peer review). **Discuss** the importance of peer review in scientific discoveries.

2.2 Variables

BY THE END OF THIS MODULE, YOU WILL BE ABLE TO:
- ✓ define variable, independent variable, dependent variable and controlled variables
- ✓ classify variables as independent, dependent or controlled.

Interactive resource
Drag and drop: Variables

GET THINKING

Working in pairs, brainstorm all of the factors that could affect how well you do in the topic test. These ideas are all variables!

Variables

A **variable** is something that can change. There may be many variables that can affect a particular outcome. For example, Figure 2.2.1 lists some variables related to the chance of successfully throwing a basketball through a hoop.

variable
a factor that could influence the result of an investigation

▲ **FIGURE 2.2.1** Variables that affect the success of throwing a basketball through a hoop

In an investigation or experiment, it is important to plan a fair test where only one variable is changed, and the others are kept the same, or controlled. This means that the investigation will test what you think you are testing, and nothing else should influence the results.

Classification of variables

Variables can be classified as independent, dependent or controlled.

- The **independent variable** is what you choose to change in your experiment.
- The **dependent variable** is what you choose to observe or measure. It may be altered by a change to the independent variable.
- **Controlled variables** are variables that you need to keep the same throughout the experiment.

independent variable
the factor that you choose to vary in your investigation

dependent variable
the factor that may be affected by the independent variable; the factor that can be measured or counted

controlled variable
a factor that needs to be kept the same throughout a scientific investigation so that it does not influence the result

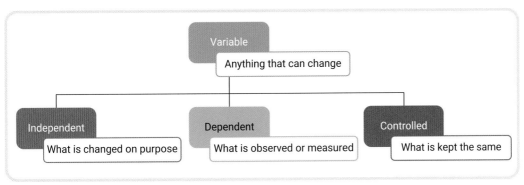

▲ FIGURE 2.2.2   Classification of variables

In the basketball example, you would choose one variable to change. For example, if you were designing a test for 'How far away from the hoop should I throw from?', the independent variable is the distance from the hoop. The dependent variable is whether the basketball goes through the hoop or not, and all the other variables would be controlled. Table 2.2.1 shows some other examples of the types of variables.

▲ FIGURE 2.2.3   The distance from the hoop is the independent variable, and so it is changed while all other variables are kept the same.

▼ TABLE 2.2.1 Examples of different variables in investigations

Type of variable	Definition	Examples		
		Does temperature affect how quickly wheat grows?	Will a heavy ball fall faster than a light ball?	Does water need to be hot to clean oily pans?
Independent	The variable that is changed on purpose	Temperature	Mass of the ball	Temperature of the water
Dependent	The variable that is measured	Rate of growth of wheat, i.e. change in height over time	Time for the ball to fall	How well the pan was cleaned
Controlled	The variables that must be kept constant	• Amount of water • Frequency of watering • Amount of soil • Size of the pot • Type of soil • Amount of fertiliser • Type of fertiliser • Amount of sunlight • Number of wheat plants • Depth of planting	• Size of the ball • Smoothness of surface of the ball • Height that the ball was dropped from • Accuracy of timer • Person timing	• Amount of water • Amount of oil on the pan • Type of oil on the pan • Type of detergent used • Amount of detergent used • Time spent cleaning • Method of cleaning/scrubbing • Size and shape of the pan

2.2 LEARNING CHECK

1 **Define**:
   a independent variable.
   b dependent variable.
   c controlled variable.

2 A toy manufacturer was testing the strength of plastic by measuring the force required to break it. **State**:
   a the independent variable.
   b the dependent variable.
   c three controlled variables.

3 Choose a question that you could test at school. **List** the independent variable, dependent variable and four controlled variables for the investigation.

4 **Explain** why a fair test only changes one variable.

5 **Create** a silly saying to remember the classification of variables. For example, 'At the zoo I changed direction to find the Dingo but Couldn't' links I (for independent) with changing, D (for dependent) with finding and C (for controlled) with couldn't.

2.3 Question, hypothesis and prediction

BY THE END OF THIS MODULE, YOU WILL BE ABLE TO:
- ✓ define prediction, hypothesis and trend
- ✓ develop a testable question from an observation
- ✓ write testable hypotheses
- ✓ identify the independent and dependent variables from a hypothesis.

GET THINKING

Think about the bees feeding on the flower shown in Figure 2.3.1.
1 What do you observe?
2 Write a question related to your observation that could be tested.
3 Write a possible answer to your question.
4 What results do you think you would get if you tested your question?

▲ FIGURE 2.3.1 Bees feeding on a flower

Quiz
Questions, hypotheses and predictions

Extra science investigation
Popcorn hypothesis

Question, hypothesis and prediction

The first two steps of the scientific method are as follows.
1 Ask a testable question about an observation.
2 Formulate a hypothesis.

Making an observation and asking a question

We are constantly making observations about the world around us. Some observations provide an awareness of our surroundings; for example, whether it is day or night. Other observations lead to actions such as putting on a jumper if it is cold.

Sometimes, an observation leads us to ask questions. For example, you may have seen chefs add salt to a saucepan of water that is being heated on a stove.

Based on this, you might make the following observation and ask a question.

Observation: Salt is added to water being heated.

Question: Will water boil more quickly if salt is added to it?

▲ FIGURE 2.3.2 Adding salt to water in a saucepan being heated on a stove

The hypothesis

From the question, you can develop a predicted explanation for the observation. This explanation is called a hypothesis. In most investigations, the hypothesis will be the prediction of how the independent variable will affect the dependent variable. The hypothesis should be based on prior knowledge or research so that it is an educated prediction and not a random guess.

A hypothesis must:
- be a statement
- be testable
- be specific for the investigation
- link the independent and dependent variables
- be based on an educated prediction, not a random guess.

In the example of adding salt to water, the independent variable is salt being added to the water and the dependent variable is how long it takes for the water to boil. All other variables, such as the size of the container, the amount of water, how the containers are heated and how you determine if the water is boiling would need to be controlled. From this, the hypothesis is what you predict adding salt to the water will do to the time it takes to boil; for example, 'If salt is added to water, then it will take less time to boil'.

Writing a hypothesis

Often a hypothesis is in the form of an 'If . . . then . . .' statement as seen in the example. A more advanced hypothesis may state how changing one factor will affect the results of another in the form of a **trend**. For example:

trend
the general direction of data; how one variable affects another

> **Question:** How does the amount of salt added to water affect the time taken to boil?
>
> **Independent variable:** The amount (mass) of salt added to the water
>
> **Dependent variable:** The time taken for the water to boil
>
> **Hypothesis:** As the mass of salt added to water increases, the time taken for the water to boil decreases.

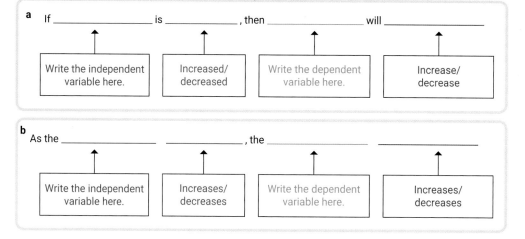

▶ FIGURE 2.3.3
(a) A simple method for writing a hypothesis.
(b) An alternative method for writing a hypothesis.

Predictions

A **prediction** is the expected result for a test. Therefore, it is what you think will happen if your hypothesis is correct.

prediction
the expected results

For example, the prediction for the salt-in-water experiment could be that the time for the water with salt to boil will be less than the time for the water without salt. Or for the more advanced investigation, the water with the most salt will boil the quickest.

ACTIVITY

For each of the images in Figure 2.3.4 make:

a an observation.
b a testable question.
c a hypothesis.
d a prediction.

▲ FIGURE 2.3.4

2.3 LEARNING CHECK

1 **Define** 'hypothesis' and 'prediction'.
2 **List** the features of a hypothesis.
3 **Classify** each of the following as an observation, a question, a hypothesis or a prediction.
   a If the temperature increases, then the number of flies decreases.
   b Ants are covering a piece of bread with honey on it.
   c The hot water will dissolve the most sugar.
   d Will the size of a discus affect how far it is thrown?
   e As the frequency of watering increases, the rate of growth of wheat increases.
   f How does the amount of fertiliser affect the growth my grandmother's roses?
4 **State** the independent and dependent variables for each of the following hypotheses.
   a As the number of days over 40°C increases, the number of flies caught in the fly trap decreases.
   b If the thickness of sticky tape increases, then the force taken to remove it increases.
   c If the lid is left off a takeaway coffee cup, then the coffee will get colder faster.
5 **Create** a flow chart, with annotations, to show the relationship between observation, question, hypothesis and prediction.
6 **Write** a hypothesis about heart rate and the intensity of exercise.
7 **Analyse** the following hypothesis by identifying ways in which it fulfils the criteria for a hypothesis and ways in which it does not fulfil the criteria: Dogs are smarter than cats.

2.4 Testing a hypothesis

BY THE END OF THIS MODULE, YOU WILL BE ABLE TO:
- ✓ define procedure, method, fair test and validity
- ✓ describe how to test a hypothesis to gain valid results
- ✓ describe how to write a repeatable method
- ✓ describe different techniques used by scientists.

Video activity
Fair testing

> **GET THINKING**
>
> A cooking recipe is a procedure that you follow. Discuss the following questions with a partner. As you work through this module, reflect on your answers and consider how they also apply to a scientific investigation.
> 1. What is the benefit of having a recipe while cooking?
> 2. Describe the outcome if everyone follows the same recipe.
> 3. What things do you need in a recipe to ensure your cooking is successful?

Planning a fair test

Any scientific investigation must be a fair test so that it is clear what has led to the results. In order to know which variable has caused the changes in the results, it is important to carefully plan a test that changes only the independent variable, and control the others. This means that you are conducting a **valid investigation** because it will provide data that tests the hypothesis.

valid investigation
an experiment that tests the hypothesis

In some investigations, it is important to include a test in which nothing is changed. This is called the **control test** and is used as a baseline to compare other data against. In the salt and water example, the control would be water without any salt.

control test
a test in an investigation in which nothing is changed

You should also collect multiple sets of data so that you can be confident that the result was not due to luck or accident. Wherever possible, conduct multiple trials, and calculate an average of the data. Other options are to complete the same experiment multiple times and for other people to conduct the same experiment.

The procedure

When planning an investigation, you write the steps that you expect to take in the **procedure**. This should include the details about how you will change the independent variable, how you will measure the dependent variable and how you will keep all the controlled variables the same. Therefore, you need to consider the following points:

procedure
a set of instructions to follow; written in the present tense

- What equipment will measure what you need? For example, you would not use a measuring cylinder to measure temperature.
- What equipment will give you accurate measurements? For example, to measure the volume of a liquid accurately, you would use a measuring cylinder, not a beaker.
- How will you test an adequate number of variations of the independent variable? For example, you could test the time for water to boil if 5, 10, 15, 20 and 25 g of salt was added to the water.
- How will you conduct sufficient trials for each test so that you can be sure that your data is reliable? For example, you could repeat each test three times and then average the data.

Method

The **method** is a set of detailed descriptions of what you did to carry out your investigation. During an investigation, scientists often need to alter the procedure. Therefore, the method is a more accurate account of how the data was collected than the planned procedure.

method
the steps that were taken during a scientific investigation; written in past tense

A good method will:

- use numbered steps
- include specific details; for example, 10 mL of water was measured with a measuring cylinder
- be clear and concise
- be written in the third person and past tense
- often be supported with a clear, labelled diagram or photo of the set-up of the experiment.

Table 2.4.1 shows the procedure and method for part of an investigation testing whether adding salt to water changes the time it takes for the water to boil.

▶ **FIGURE 2.4.1** Testing the hypothesis that adding salt to water makes it boil faster

▼ **TABLE 2.4.1** Part of the procedure and method for a salt and water investigation

	Procedure	Method
1	Measure 100 mL of water and pour it into a beaker.	100 mL of water was measured with a measuring cylinder and poured into a 250 mL beaker.
2	Light the Bunsen burner and open the air hole to change the flame to a blue flame.	The Bunsen burner was lit and then the air hole was opened to change the flame to a blue flame.
3	Use tongs to place the beaker on a gauze mat on a tripod stand over a Bunsen burner.	The beaker was lifted using tongs and placed on a gauze mat on a tripod stand over a Bunsen burner.
4	Start the stopwatch as soon as the beaker is over the Bunsen burner.	The stopwatch was started as soon as the beaker was over the Bunsen burner.
5	Stop the stopwatch when the water is boiling.	The stopwatch was stopped when the water was boiling.
6	Repeat steps 1–5 four times.	Steps 1–5 were repeated four times.

Techniques used to collect data

Scientists collect data using different techniques. For example:

- A survey could collect data about people's opinions.
- Field studies could collect data about the environment.
- Sampling could collect data about wildlife.
- Laboratory tests could collect data about chemicals.
- Observations could collect data about behaviour.
- Clinical trials could collect data about new drugs.

▲ **FIGURE 2.4.2** One way scientists collect data is as a field study in the environment.

ACTIVITY

How accurate is your method?

Aim

To work in groups of three to test the accuracy of each other's method

You need

- paper
- coloured pens or pencils
- ruler
- 20 mL measuring cylinder
- 100 mL beaker
- food dye
- stirring rod
- items in a pencil case
- device to record a video (if possible)

What to do

1. Allocate the following roles in the group.
   - Person A completes a task initially.
   - Person B writes the method.
   - Person C completes the task a second time.
2. Person C moves to an area where they cannot see what persons A and B are doing.
3. Person A completes one of the following tasks. If possible, record them doing this on a video.
   a. Draw a simple shape with different coloured pens on a piece of paper.
   b. Pour exactly 20 mL of water and 3 drops of food dye into a 100 mL beaker and then stir the solution.
   c. Arrange the items from a pencil case into a certain design.
4. Person B writes a method for what person A did without talking to person A.
5. Put the finished product from step 3 out of sight.
6. Person C returns to the group and, using only the method, does the same task as person A.
7. Compare the finished products from persons A and C. How accurate was person B's method?
8. Swap roles and repeat the process so that everyone has a go at each role.

What do you think?

1. What characteristics did the method need to have for person C to closely recreate person A's product?
2. What challenges were there in writing an accurate method?

2.4 LEARNING CHECK

1. **List** four features of a procedure for a fair test.
2. **Explain** why a valid test only changes one variable.
3. Write a procedure and method for brushing your teeth.
4. **Explain** why it is important for a method to be detailed.
5. Use the following brief procedure to **write** a method for testing the bounciness of different balls.
   1. Drop the ball from 1 m high.
   2. Count how many times the ball bounces before it stops.
   3. Repeat this for other balls.
   4. Test all balls three times.
6. **Explain** why a method is written in the past tense.

2.5 Recording results

BY THE END OF THIS MODULE, YOU WILL BE ABLE TO:
- ✓ describe the components of an effective table
- ✓ organise data in an effective table
- ✓ calculate the mean of a set of data.

Interactive resource
Match: Data and results

GET THINKING

The information about your classes for each day is organised into a timetable.

1. Describe how the information is arranged in your timetable.
2. Explain the benefits of having the information arranged in a timetable.
3. Predict what would happen if you didn't have a timetable.

▲ FIGURE 2.5.1 A school timetable

data
the numbers or observations collected during an experiment

result
the information gained from an experiment

row
a horizontal division in a table

column
a vertical division in a table

anomaly
something that deviates from the standard

Recording results

When scientists conduct an investigation, they make observations and measurements. This is called **data**. The information gained from this data is the **result** of the investigation.

Organising data

It is important to organise and evaluate the data correctly so that valid conclusions are made. The most common, effective way to organise data is in a table, with **rows** and **columns** (e.g. Figure 2.5.3). This makes it easier to identify trends in the results and any **anomalies** in the data.

Figure 2.5.4 on the next page shows the features of an effective table.

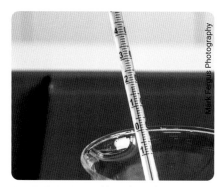

▲ FIGURE 2.5.2 How does the temperature of the water affect how long it takes for a dissolvable tablet to dissolve?

Table 1. The time for one dissolvable tablet to dissolve in water of different temperatures

Temperature of 100 mL of water (°C)	Time for one dissolvable tablet to dissolve (s)			
	Trial 1	Trial 2	Trial 3	Average
5	320	345	330	331.7
25	283	266	250	266.3
70	25	30	28	27.7

▲ FIGURE 2.5.3 A table showing the data collected from an investigation of the time taken for a dissolvable tablet to dissolve in water at different temperatures. The data provides the results of the investigation.

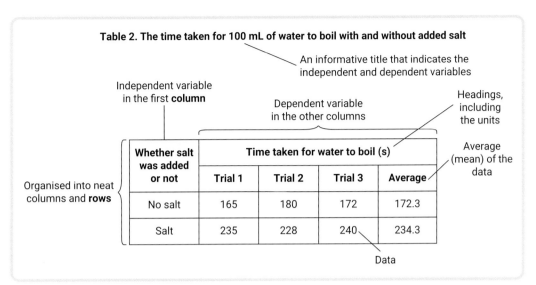

▲ FIGURE 2.5.4　Features of an effective table

Finding an average

For some investigations, you will collect a large amount of quantitative data. In many cases, you will want to find an average value, or **mean**. To find the mean:

- add together all the data for one variable to obtain a total
- divide the total by the number of pieces of data.

mean
the calculated 'central' value of a set of number; an average

For example, you may want to find the average height of students in your class.

The data in Table 2.5.1 show the heights of a group of students. Adding all the data gives:

$$163 + 158 + 172 + 180 + 163 + 150 + 168 + 175 + 155 = 1484$$

Dividing by the number of pieces of data (nine students) gives:

$$\text{Average height} = \frac{\text{total of heights}}{\text{number of students}}$$
$$= \frac{1484}{9}$$
$$= 164.8$$
$$= 165$$

Therefore, the average height of the students is 165 cm.

▼ TABLE 2.5.1　Heights of students

Student	Height (cm)
Alex	163
Brianna	158
Cam	172
Daniel	180
Dora	163
Ellie	150
Huang	168
John	175
Milly	155

2.5 LEARNING CHECK

1 **List** the features of an effective table.
2 **Explain** why an effective table for results is beneficial during an investigation.
3 **Calculate** the mean of each of the following sets of data.
   a 4.2, 8.1, 5.33, 9.7
   b 140, 125, 177, 139, 121, 106
   c 0.08, 0.066, 0.09
   d 10 500, 13 008, 11 345, 12 900, 15 045
4 During an investigation, Ajang measured the height of grass that was planted in different types of soil (sand, potting mix, compost, gravel and clay). **Construct** a table for Ajang to record his results in.

2.6 Analysing results

BY THE END OF THIS MODULE, YOU WILL BE ABLE TO:
- ✓ represent data in appropriate graphs
- ✓ analyse data to identify patterns and trends.

Interactive resource
Drag and drop: Types of data

GET THINKING

You may have already seen, or used, different types of graphs to represent information. List the types of graphs that you know, sketch what they look like and state a situation where they may be used.

Types of graphs

pattern
when data repeats in a predictable manner

discrete data
data where there is only a limited number of possibilities

continuous data
measurements on an infinite scale so any value between two numbers is possible

Data can be represented visually in a graph. This makes it easier to see **patterns** or trends in the data.

The type of graph used is determined by the type of data that is recorded. **Discrete data** can only have certain values, whereas **continuous data** can have infinite values between two numbers. For example, the number of students in a class is discrete data because you cannot have half a person. By contrast, student height is continuous data because it is possible to have a range of values as a height.

Two of the most common types of graphs in science are column graphs and line graphs. Column graphs, such as the one in Figure 2.6.1, are used to represent discrete data and qualitative data. Line graphs, such as the one in Figure 2.6.2 and Figure 2.6.3, are used when both variables are continuous data.

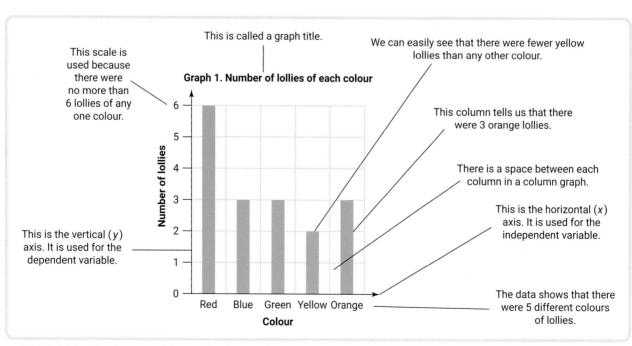

▲ **FIGURE 2.6.1** A column graph is used for discrete data.

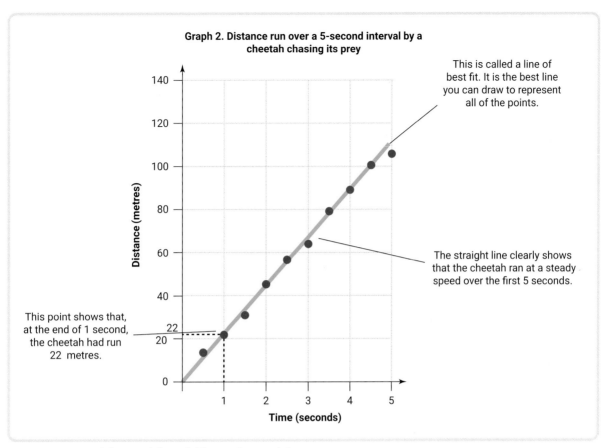

▲ FIGURE 2.6.2   A line graph is useful for continuous data.

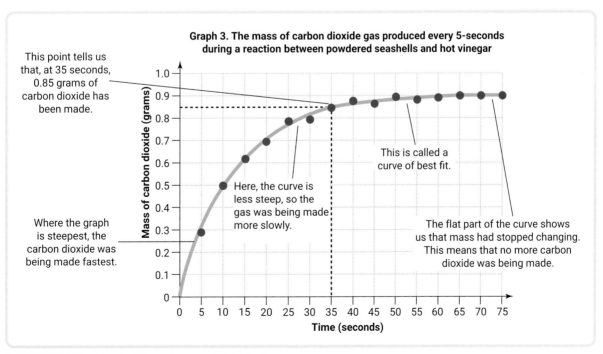

▲ FIGURE 2.6.3   How to interpret a curved line graph

Drawing graphs

While the type of graph may vary, there are some similarities between them.

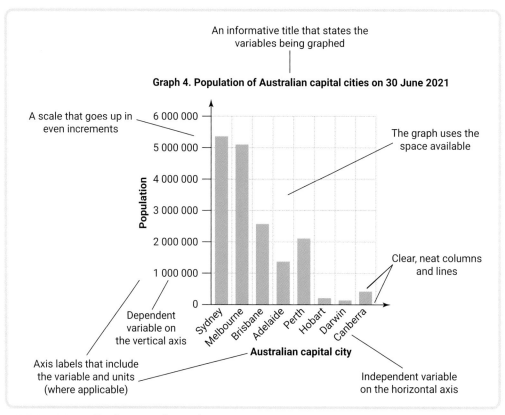

▲ FIGURE 2.6.4   Key features of a graph

1. An informative title that includes the independent and dependent variables
2. Clearly labelled, straight horizontal and vertical axes
3. The independent variable graphed on the x-axis (the horizontal axis)
4. The dependent variable graphed on the y-axis (the vertical axis)
5. A label on each axis that states the variable
6. Units in brackets (if appropriate) with the axis label
7. A scale on each axis that goes up in even increments if the data is numerical
8. In most cases, only the average of the data is graphed
9. A key to identify data if there is more than one set of data graphed on one set of axes
10. A line of best fit wherever possible for line graphs
11. Using the maximum area of the graph paper (within reason) when hand drawn

Analysing data

Once the data is organised, it is analysed for patterns and trends. A pattern is observed when the data repeats in a predictable way. For example, an **analysis** of the way in which the average temperatures decrease and then increase during each year reveals a pattern. A trend, or relationship, between the variables is described when the dependent variable

analysis
the careful study of data to look for patterns and trends

consistently changes in a certain way. For example, an increasing trend exists between the mass of the object and the energy used to lift an object. Figures 2.6.5 and 2.6.6 show examples of different trends between variables.

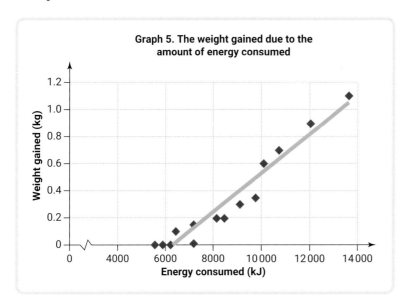

◀ FIGURE 2.6.5 Analysing this graph shows a trend of increasing weight gain as the number of kilojoules consumed increases.

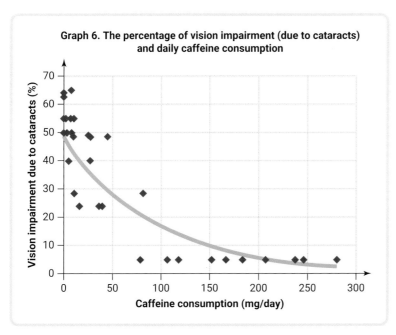

◀ FIGURE 2.6.6 Analysing this graph shows a trend of decreasing vision impairment due to cataracts as caffeine consumption increases.

If the trend is the same as the predicted trend, then the results support the hypothesis. However, they do not prove the hypothesis because many tests need to be done before scientists can be sure that the relationship is always the same.

The hypothesis for an investigation should be based on previous information. Therefore, the trend in the results can be explained using scientific knowledge.

2.6 LEARNING CHECK

1 Copy and **complete** the table to classify data as discrete or continuous.

Data	Continuous or discrete?
Eye colour	
Rainfall	
Favourite food	
Temperature	
Time	

2 **Compare** continuous and discrete data.

3 What type of graph should be used for:
   a the length of a day at different times of year?
   b the volume of carbon dioxide released from different types of plants?
   c the mass of a chicken every day over 3 months?
   d whether people prefer coffee, tea or hot chocolate?

4 Table 2.6.1 shows the results of an experiment for the time taken for bleach to remove the colour from dye at different temperatures.

▼ TABLE 2.6.1 The time for bleach to remove the colour of dye at different temperatures

Temperature (°C)	Time taken for the colour of dye to disappear (seconds)			
	Trial 1	Trial 2	Trial 3	Average
0	240	235	242	
20	189	185	186	
50	80	75	78	
70	20	22	21	

   a **State** the independent variable and the dependent variable.
   b **Write** a possible hypothesis for the investigation.
   c **Calculate** the average of the times at each temperature.
   d **Graph** the average time against temperature.
   e **Describe** the trend between the temperature of the bleach and the time for the dye to lose its colour.

5 **Describe** the trend shown in each of the following graphs.

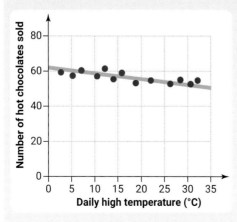
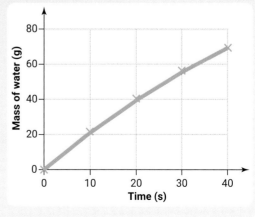

2.7 Evaluation and conclusion

BY THE END OF THIS MODULE, YOU WILL BE ABLE TO:
- ✓ define accurate, reliable, repeatable, reproducible and valid
- ✓ evaluate data for accuracy, reliability and repeatability
- ✓ evaluate an investigation for validity and reproducibility
- ✓ write a valid conclusion.

Interactive resource
Crossword: The scientific method

GET THINKING

Look at Figure 2.7.1, which shows the results when darts were thrown at a dartboard, with the aim of hitting the red circle in the centre. Use the image to predict what is meant by accuracy and reliability of data.

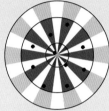
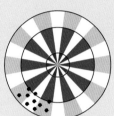
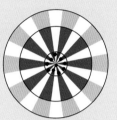

High accuracy, low reliability | Low accuracy, high reliability | High accuracy and high reliability

▲ FIGURE 2.7.1 The accuracy and reliability of throwing darts

Evaluation

The next step of the scientific method is to evaluate the data and the method. Evaluation allows scientists to understand how trustworthy any conclusions made during an investigation are.

For a thorough evaluation, the investigation needs to be repeated by the same person as well as by other people. This provides multiple, independent results that can be compared.

Table 2.7.1 summarises the information to discuss in an evaluation.

▼ TABLE 2.7.1 Factors to consider when evaluating a scientific investigation

Characteristic	Definition	Factors to consider in the investigation
Accuracy of the data	How close a measurement is to the correct value	• Choosing the correct equipment • Choosing the most appropriate size of equipment • Using the equipment correctly
Reliability of the data	How similar data is when collected from repeated experiments	• Can only be identified if multiple trials or tests are conducted • Increased by a clear, detailed method • Increased by using exactly the same method each time
Repeatability	How similar data is when collected by the same person conducting the same experiment	
Reproducibility	How similar data is when collected by different people conducting the same experiment	
Validity of the method	The extent to which an investigation tests the hypothesis	• The independent variable is changed • The dependent variable is measured • All other variables are controlled

accuracy
how close a measurement is to the correct value

reliability
how similar the results of the same experiment are

repeatable
the same results are obtained when the same person conducts the same experiment

reproducible
the same results are obtained when a different person conducts the same experiment

validity
the extent to which an investigation tests the hypothesis

Conclusion

conclusion
a judgement reached by reasoning

The **conclusion** is a clear statement of what the investigation found and whether or not it supported the hypothesis. The conclusion should not introduce any new information or analysis; it is simply summing up the findings.

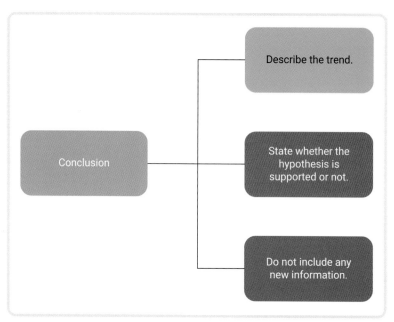

▲ FIGURE 2.7.2   A summary of what to include in a conclusion

2.7 LEARNING CHECK

1 **Define**:
   a accuracy.
   b reliability.
   c validity.
2 What is the difference between the repeatability and the reproducibility of data?
3 Reflect on the hypothesis and trend identified in Learning check 2.6 Question 4. **Write** a conclusion for this information.
4 **List** two things that would reduce the accuracy of an investigation that was measuring the force required to tear paper of different thicknesses.
5 **Explain** how a clear method allows reliable data to be collected in an investigation.
6 Before a new pharmaceutical drug is released to the public, it must undergo several tests. **Discuss** why it is important that reproducible results are obtained before the drug is made available to the public.

2.8 An example of a scientific report: Growing tomato plants with fertiliser

BY THE END OF THIS MODULE, YOU WILL BE ABLE TO:
- ✓ list the parts of a scientific report
- ✓ write a scientific report.

GET THINKING

This module introduces you to writing a scientific report to share the findings of an investigation. As you work through the module, summarise how to write each section in a flow chart, as in Figure 2.8.1.

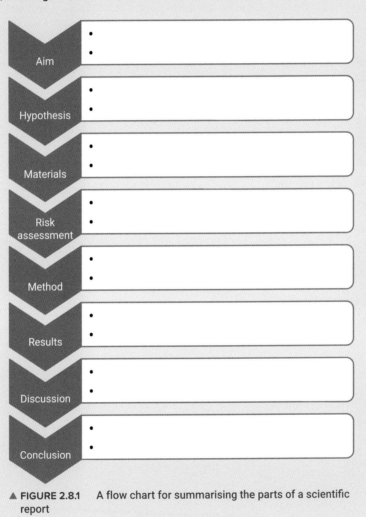

▲ FIGURE 2.8.1 A flow chart for summarising the parts of a scientific report

Interactive resource
Label: Parts of a scientific report

Science report

Science reports are used to communicate information about scientific investigations. Although the format of the report may vary depending on the purpose and audience, the most common forms are a formal report or a poster. Science reports usually include the sections shown in the following report for an investigation on tomato plants.

Aim

> State the purpose of the investigation.

To investigate the effect of fertiliser on the rate of growth of tomato plants

Hypothesis

> State the hypothesis.

If fertiliser is applied to the soil, a tomato plant will grow faster than if no fertiliser is added.

Materials

> List, in detail, the equipment and materials needed.

- 6 tomato seedlings
- 6 small rectangular pots, each measuring 4 cm × 4 cm × 6 cm
- potting mix
- 25 mL measuring cylinder
- water
- fertiliser (Aquasol)
- electronic balance
- 30 cm ruler (showing millimetres)

Risk assessment

> State any hazards and how the risk can be managed.

Potting mix contains fungal spores that could be harmful to health. Use potting mix in a well-ventilated area and wear a dust mask and gardening gloves.

Method

> Write a detailed method as a series of numbered steps and a labelled diagram or photo.

1. The pots were numbered as 1–6.
2. Each pot was filled with potting mix to a height of 3 cm.
3. One tomato seedling was planted in the centre of each pot so that the potting mix was at the same level as the soil the seedling was in.
4. The heights of the seedlings were measured with a ruler.
5. Pots 1–3 were labelled 'Water only'.
6. 10 g of Aquasol fertiliser was sprinkled onto the soil of pots 4–6. These pots were labelled 'Fertiliser'.
7. Each pot was watered with 20 mL of water at the same time each day.

8 The heights of the seedlings were measured every second day at the same time, where possible.
9 Steps 7 and 8 were repeated for 2 weeks.

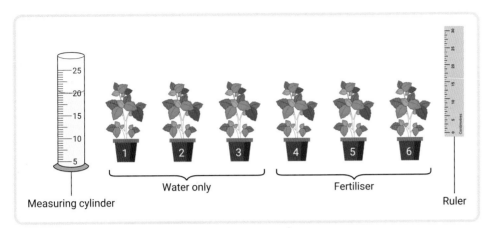

▲ FIGURE 2.8.2 The set-up of the experiment

Results

Tables

Include a table of results and any observations.

▼ TABLE 2.8.1 The heights of tomato plants with and without fertiliser over a 14-day period

Day	Heights of tomato plants without fertiliser (cm)			Heights of tomato plants with fertiliser (cm)		
	Pot 1	Pot 2	Pot 3	Pot 4	Pot 5	Pot 6
0	17.5	18.5	21	20	19	18
2	19	20	21	21	21.5	20.5
4	22.5	23.5	23	22	24	23
7	25	28	27	26.5	28	26.5
9	27.5	28.5	28	28.5	30.5	28
11	29	29	29	33	33.5	32.5
14	32	33.5	36.5	36	39.5	38.5

▼ TABLE 2.8.2 The increase in heights of tomato plants with and without fertiliser

With or without fertiliser	Increase in heights of tomato plants over 14 days (cm)			
	Trial 1	Trial 2	Trial 3	Average
No fertiliser	14.5	15	15.5	15
With fertiliser	16	20.5	20.5	19

Graphs

Include a graph of the results.

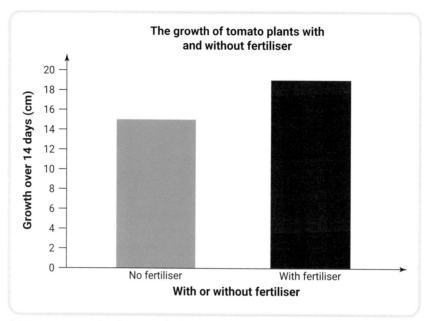

▲ FIGURE 2.8.3   A column graph showing the increase in height of tomato plants over 14 days with and without fertiliser

Discussion

Discuss the analysis and evaluation of the investigation, including:
- the trend found, the scientific reason for it and whether it supports the hypothesis
- the accuracy, reliability and repeatability of the data, including any anomalies
- the validity of the method
- ways that the investigation could be improved or extended.

Six tomato seedlings were grown under identical conditions, except that 10 g of Aquasol fertiliser was added to the soil of three plants. Results showed that the tomato plants that received the fertiliser grew, on average, 4 cm more than the ones that did not receive fertiliser. At weekends, the plants did not receive any water. Because this affected all plants equally, the difference in growth would not have been affected.

Both sets of plants initially grew at very similar rates until day 9, after which the seedlings with the fertiliser grew more quickly than those without fertiliser.

These results support the hypothesis that a tomato plant in soil with fertiliser grows faster than one in soil without fertiliser. This is because the addition of the fertiliser provided the tomato seedlings with additional nutrients, which helped them to grow faster.

The method fairly tested the hypothesis because all variables were controlled with the only variation being the addition of Aquasol fertiliser to half the pots. The method was also appropriate because it measured the height of the tomato plants, which is a valid method of measuring the growth of the plants. This allowed the growth of the plants with and without fertiliser to be compared.

Only the height of the plants was measured. This did not consider the width that the plants grew. Therefore, the accuracy of the growth was limited.

The data for the plants without fertiliser was very reliable, with the growth varying by only 1 cm. The data for the plants with fertiliser was less reliable, with pot 4 growing 4 cm less than the other pots. Overall, the results are repeatable because the results are similar and each trial conformed to the same trend.

Any future investigations of this type should include more plants under each condition. This would provide more data and minimise problems arising if any of the plants die during the experiment. The investigation could also be extended to test the effect of other fertilisers, the effect of different amounts of fertiliser, the effect of the frequency of adding the fertiliser or the effect of fertiliser on the growth of different plants.

Conclusion

> State the trend and whether it supports the hypothesis.

Results from this experiment support the hypothesis that tomato plants with fertiliser added to their soil grow faster than ones without fertiliser added.

2.8 LEARNING CHECK

1 In the tomato plant investigation, what do we call the plants that did not have fertiliser added to their soil? What is the purpose of these tomato plants?
2 If you used a different type of potting mix for each tomato plant, what effect might that have on the investigation?
3 **Describe** the implications of one of the tomato plants dying during the investigation.
4 **Explain** why it is important for the method to be clear, detailed and thorough.
5 Can you say conclusively that all tomato plants will grow faster if fertiliser is added? **Explain** your answer.

2.9 Scientists as communicators

SCIENCE AS A HUMAN ENDEAVOUR

BY THE END OF THIS MODULE, YOU WILL BE ABLE TO:
- ✓ explain why it is important for scientists to communicate their findings.

Communicating is an important part of the work of scientists. It allows scientists to have their investigations peer-reviewed and reproduced, source funding for their work, collaborate with other scientists and share what they learn with the wider community and people who need to use their findings. Dame Jane Goodall is an example of a scientist who is an effective communicator.

Dame Jane Goodall is well known, not only among scientists, but also by the general public worldwide. In 1957, at the age of 23, Jane began her study of chimpanzees in Africa. By living in the chimpanzees' habitat, she could observe behaviours that hadn't previously been seen, such as eating meat and using tools. These observations changed our understanding of chimpanzees, and our knowledge about the relationships between primates.

In 1963, Jane published an article in *National Geographic* titled 'My Life Among Wild Chimpanzees'. Since then, she has written many articles and more than 25 books. She has inspired many films and has received many awards. Jane Goodall now spends her time travelling to speak about the threats to chimpanzees and the environment, urging audiences to recognise their power to change the world for the better.

The impact that Jane Goodall has had was only possible because she shared her experiences and findings with the wider community. Jane has engaged specific audiences through different communication methods. By adjusting her communication from scientific reports to children's books to popular movies, Jane has been able to share her studies with a lot of people.

▲ **FIGURE 2.9.1** Jane Goodall communicated the results of her studies about the behaviour of chimpanzees in Tanzania.

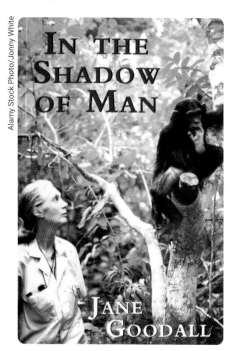

▲ **FIGURE 2.9.2** One of the books that Jane Goodall has written

2.9 LEARNING CHECK

1. **List** the benefits of scientists communicating with different audiences.
2. **Explain** how Jane Goodall has managed to use her studies to promote environmental conservation.
3. **Describe** the ways that scientists can communicate with the general public.
4. Choose one other scientist and **discuss** how they have communicated to drive change.

Video activity
Science communication tips

SCIENCE INVESTIGATIONS

2.10 Writing hypotheses

SCIENCE SKILLS IN FOCUS

IN THIS MODULE, YOU WILL FOCUS ON LEARNING AND IMPROVING THESE SKILLS:

- Writing a hypothesis

A hypothesis is a testable explanation for something based on existing knowledge. It makes a link between what is changed and the outcome that is being tested.

When writing a hypothesis:

1 Identify the dependent variable – what is being measured or observed?
2 Identify the independent variable – what is being changed?
3 How do you think the independent variable will affect the dependent variable?

A common way to write a hypothesis is with an 'If …, then …' statement, as seen in Figure 2.10.1.

For example, a question being tested is 'Does the thickness of paper change how far a paper plane flies?' Then the hypothesis might be 'If the thickness of the paper is increased, then the distance that the plane flies will decrease'.

Video
Science skills in a minute: Hypotheses

Science skill resource
Science skills in practice: Writing hypotheses

HOW QUICKLY DOES BREAD GO MOULDY?

AIM

To investigate how temperature affects the time it takes for mould to grow on bread

WHAT TO DO

1 Work in groups of three or four.
2 List the independent, dependent and controlled variables.
3 Write a hypothesis stating the predicted effect of temperature on the time it takes for mould to grow on bread (see Figures 2.10.1 and 2.10.2.).
4 Develop a plan for testing the hypothesis, including:
   a the temperatures you will test.
   b how you will identify when mould first appears.
   c how many pieces of bread you will test at each temperature.
   d how you will keep other variables (e.g. light, type of bread and moisture) the same for all tests.
5 Write a detailed list of materials for the investigation.
6 Conduct the investigation and record your results.

 Warning

Mould spores can irritate the lungs. Place bread in sealed plastic bags at the start of the investigation. Always keep the bags sealed, including when disposing of them at the end of the investigation.

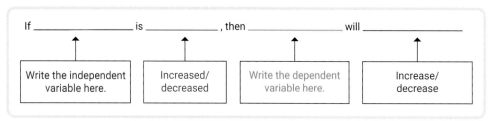

▲ FIGURE 2.10.1 A simple method to write a hypothesis

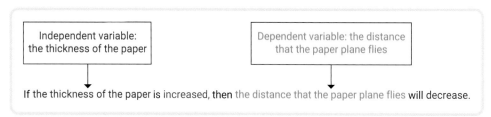

▲ FIGURE 2.10.2 An example of a hypothesis

RESULTS

Use the following template to create a table for your results.

▼ Results table *Add the title here*

Add the heading (and units) for the independent variable here	Add the heading (and units) for the dependent variable here			
	Trial 1	Trial 2	Trial 3	Average

WHAT DO YOU THINK?

Analyse the results

1. How did the temperature affect how long it took for mould to grow on the bread?
2. Suggest a reason for the relationship between temperature and the growth of mould on bread.

Evaluate the results

3. Discus whether the data was accurate.
4. Discuss whether the data was repeatable (in other words, whether the data was consistent between trials).

Evaluate the method

5. Discuss whether the method was valid by considering if it tested the hypothesis.
6. Discuss whether the results are valid by considering if the variables were controlled.

CONCLUSION

Summarise the trend in the results and whether it supports the hypothesis or not.

2 REVIEW

REMEMBERING

1 **List** the steps, in order, of the scientific method.
2 What factors must you consider to make an investigation a fair test?
3 **State** which variable is shown on the:
   a   vertical axis of a graph.
   b   horizontal axis of a graph.
4 What are two different types of graphs? When are they used?

UNDERSTANDING

5 **Explain** why it is important to only change one variable in an investigation.
6 **Outline** the difference between a method and a procedure. Use this to **explain** why a method is written in past tense.
7 **Explain** why we do not say an experiment proves a hypothesis correct.
8 Why is it important that scientific investigations are repeatable?
9 **Explain** why it is important for your materials list to be detailed.

APPLYING

10 The hypothesis for an experiment was 'As the mass of a single ice cube increases, the time taken to melt will increase'. **State** the:
   a   independent variable.
   b   dependent variable.
   c   four controlled variables.

EVALUATING

11 Consider the data in Table 1 that was collected during an investigation of the temperature during a chemical reaction between vinegar and sodium bicarbonate (baking soda).
   a   **Classify** the data as continuous or discrete.
   b   **Graph** the data.

▼ TABLE 1 The temperature as the reaction between vinegar and sodium bicarbonate progresses

Time (s)	Temperature (°C)
0	21.0
5	20.5
10	19.0
15	17.5
20	16.8
25	16.5
30	16.1
35	16.0

   c   **Describe** the trend shown in the graph.
   d   **Predict** the temperature at 12 seconds.

12 An activity in your science class required you to measure the temperature in your science classroom over a 4-hour period. Your results are shown in Table 2.
   a   Is this data described as continuous or discrete? **Justify** your answer.
   b   **Represent** this data as a graph.
   c   What was the highest temperature during the 4-hour period?
   d   **Describe** the trend shown in your graph.
   e   From your graph, determine the temperature of the room at 45 minutes.

▼ TABLE 2 The temperature of the classroom over a 4-hour period

Time (min)	Temperature (°C)
0	14
30	16
60	19
90	21
120	21
180	23
240	17

13 During an experiment, the mass of limestone that dissolved was measured. The results were 5, 7.2, 6.3, 5.5, 5.8, 6, 4.9 and 6.7 g. **Calculate** the mean of this set of data.

14 Olivia and Ethan decided to test whether red flowers lasted longer than white flowers. They put a red rose in a small cup of water on the bathroom shelf, and a white carnation in a large jar of water on the bench in the garden shed. After 3 days, the white flower had died, but the red flower was still alive. They concluded that red flowers lasted longer than white flowers.
   a **Explain** why this was not a fair test.
   b Was Olivia and Ethan's conclusion valid?
   c **Rewrite** their method to make this a fair test.

15 In your opinion, what might happen if every scientist reported their scientific discoveries in a different way?

16 Consider the information represented in graph 1.
   a What type of graph is shown?
   b **State** the independent variable.
   c **State** the dependent variable.
   d **List** three variables that would have been controlled in this investigation.
   e **Describe** the trend shown in the graph.

17 A scientist claimed to have made an important discovery but did not follow the scientific method or write up a scientific report. What are the implications of this? Would that scientist's findings be taken seriously?

CREATING

18 **Write** a hypothesis for each question.
   a Why do my black clothes get hotter in summer than my white clothes?
   b What would make my model car go faster?
   c Will plants grow more if I give them more water?
   d How does the type of mug affect how quickly a cup of coffee goes cold?
   e Does ice melt faster if left out on a tray or put in room temperature water?
   f Do diet drinks help you lose weight?

19 **Plan** and **conduct** an investigation to test the question 'Do people prefer to read novels as a hard copy or on a device?' **Create** a scientific report to share your findings.

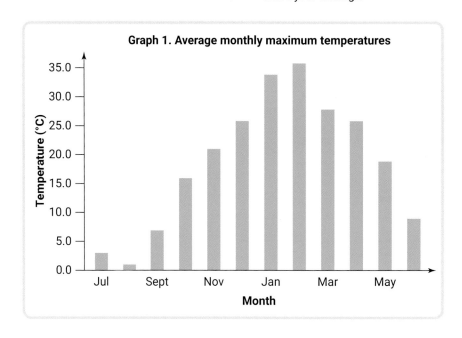

Graph 1. Average monthly maximum temperatures

BIG SCIENCE CHALLENGE PROJECT #2

What questions do you want answered?

1. Connect what you've learned

Scientists progress their understanding of the world by continually asking questions and then using the scientific method to find answers that can be trusted. List five questions that you could test to find the answers. For example, do the more expensive nappies keep babies drier? Or, how much water does the lawn need in summer?

2. Check your thinking

Choose a question that can be safely tested at school.

- What will you find out? This is your dependent variable.
- Write a list of everything that could affect the results (the variables).
- Choose one variable to test – this will be your independent variable.
- Write a hypothesis to describe how you think the independent variable will affect the dependent variable.
- Decide how you will keep all other variables constant.

3. Make an action plan

Create a procedure to test your hypothesis, including a list of materials and a risk assessment.

Ask your teacher to check your procedure. With their permission, carry out your test.

4. Communicate

Create a poster to share what you did and what you found. Use the following headings to organise the information so that it is clear for your audience.

- Aim
- Hypothesis
- Materials
- Risk assessment
- Method
- Results
- Discussion
- Conclusion

3 Matter: solids, liquids and gases

3.1 Matter (p. 70)
Matter is anything that has a mass and occupies space.

3.2 Particle theory of matter (p. 72)
Matter is made up of particles.

3.3 States of matter (p. 74)
Matter can exist as a solid, a liquid or a gas.

3.4 Properties of solids (p. 78)
The properties of a solid are due to the particles being held closely together and having limited movement.

3.5 Properties of liquids (p. 80)
The properties of a liquid are due to the particles being able to move past one another.

3.6 Properties of gases (p. 82)
The properties of a gas are due to the particles being as far apart as possible and moving very quickly.

3.7 Changing state (p. 84)
Matter will change state if enough energy is added or removed.

3.8 SCIENCE AS A HUMAN ENDEAVOUR: Collecting drinking water (p. 88)
Drinking water can be collected from the air.

3.9 SCIENCE INVESTIGATIONS: Organising and representing data (p. 90)
1. Melting and boiling points of water
2. Melting points of matter

BIG SCIENCE CHALLENGE #3

▲ FIGURE 3.0.1 Playing sport can make you hot and sweaty.

When you play sport or stand out in the sun on a hot day, you sweat. At the time, this seems annoying; however, it is lifesaving because it helps your body cool down so that it doesn't overheat.

Some people and animals have a condition called anhidrosis. This means they are unable to sweat. Anhidrosis can be life-threatening because people with this condition cannot cool themselves down and are more prone to heatstroke.

▶ How do you think sweating cools you down?

▶ What could people with anhidrosis do to help cool themselves?

 SCIENCE CHALLENGE ACCEPTED!

At the end of this chapter, you can complete Big Science Challenge Project #3. You can use the information you learn in this chapter to complete the project.

Assessments
- Prior knowledge quiz
- Chapter review questions
- End-of-chapter test
- Portfolio assessment task: Science investigation

Videos
- Science skills in a minute: Organising data **(3.9)**
- Video activities: Solids, liquids and gases **(3.3)**; Properties of gases **(3.6)**; Changing states of matter **(3.7)**

Science skill resources
- Science skills in practice: Organising and representing data **(3.9)**
- Extra science investigations: Particle movement in liquids **(3.5)**; Changes of state and the weather **(3.7)**

Interactive resources
- Simulations: Introduction to gases **(3.6)**; Phase changes **(3.7)**
- Drag and drop: Describing matter **(3.1)**; Solids v liquids **(3.5)**
- Crossword: Changing states **(3.7)**

Nelson MindTap

To access these resources and many more, visit
cengage.com.au/nelsonmindtap

Chapter 3 | Matter: solids, liquids and gases

3.1 Matter

BY THE END OF THIS MODULE, YOU WILL BE ABLE TO:
✓ describe matter, including different examples.

> **GET THINKING**
>
> We take the things around us for granted, and don't think about what they are made of. Look around you. Can you think of a term that describes everything?

Interactive resource
Drag and drop: Describing matter

What is matter?

When you look around you, it is easy to identify a wide range of different substances. If you were to list the substances within reach while you sit in your classroom, you could fill a page. There may be the top of the desk, paper, iPad, pen, water, other students, the air, the plastic of the seat, the metal of the frame of the desk or chair, the carpet … and the list goes on! Although it may appear that all these things are different, they can actually all be classified as **matter**.

matter
anything that takes up space and has mass

▲ **FIGURE 3.1.1** Matter is all around you.

mass
the amount of matter in an object, measured in kilograms (kg), grams (g) or milligrams (mg)

space
the three-dimensional region (length, width and height) where an object exists

Matter is defined as anything that has **mass** and occupies **space**. That sounds simple, but what does it actually mean?

Mass is a measure of the amount of substance, or matter, that makes up an object. In Chapter 1, you learned that we usually use an electronic balance to measure mass, and that it is measured in kilograms (kg), grams (g) or milligrams (mg). If we weighed a tennis ball and a baseball, the baseball would have a greater mass because it is made up of more matter.

Space is a three-dimensional region where things can exist. In other words, if something has a height, width and length, then it is occupying space.

▲ **FIGURE 3.1.2** The baseball has a greater mass than the tennis ball because it is made up of more matter.

Describing matter

As scientists, it is important that you can describe matter in a such way that other scientists will be able to identify the object. Therefore, we use **properties** that are consistent and clear so that there are no misunderstandings.

property
a characteristic or feature of a substance

A property is a characteristic used to describe something by observation and/or measurement. Some common properties are colour, shape, texture, smell, flexibility, hardness, strength and dimensions.

The clearer and more specific the properties are, the more accurate the description will be. For example, describing something as a green solid could be confusing because the object could be broccoli, grass, a green box or a tennis ball. However, if you described it as a green, spherical solid of approximately 6.5 cm diameter with a soft, uneven texture, that could be squashed, and when it was dropped it returned to a similar height, then it is very likely to be a tennis ball.

3.1 LEARNING CHECK

1 **List** the two properties exhibited by all matter.
2 **Explain** why a truck has a greater mass than a car.
3 Choose an item out of your pencil case and write a **description** using only its properties. Swap descriptions with a partner and see if you can identify each other's object.
4 **Suggest** some problems that might occur if scientists did not have a common language and understanding when describing matter.
5 Is air classified as matter? **Justify** your answer.

3.2 Particle theory of matter

BY THE END OF THIS MODULE, YOU WILL BE ABLE TO:
- ✓ state the particle theory of matter
- ✓ describe the structure of matter by applying the particle theory of matter.

GET THINKING

After reading this module, you will be asked to explain how the structure of matter is similar to students in a school. As you read the information, think about how the two things are similar.

Quiz
What's the particle theory of matter?

Particle theory of matter

For a long time, scientists have tried to explain what matter is made of. As early as 2500 years ago, Greek philosopher Democritus developed a theory that stated that the universe consisted of empty space and an infinite number of particles. These particles were different from each other in form, position and arrangement. In the 19th century, English chemist John Dalton concluded that all matter was composed of a single, unique type of particle and that these particles could not be divided into smaller particles.

We now know that Democritus and Dalton were both right. Even though we can't see them, everything is made up of **particles**. For example, a cube of sugar might look like just one structure, but if you looked closely, you would see that it is made of smaller grains of sugar. If you could zoom in on the grains, you would see that they are made up of smaller units, which we call particles (Figure 3.2.2).

▲ **FIGURE 3.2.1** Ancient Greek philosopher Democritus developed a theory about particles.

particle
a tiny unit of matter

model
a simplified explanation that makes something easier to explain or understand

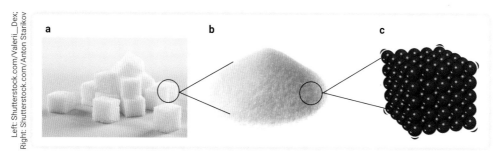

▲ **FIGURE 3.2.2** (a) A sugar cube is made of (b) many grains of sugar, which are made of (c) particles.

Particles are too small for us to see, even with the most powerful electron microscope. Therefore, scientists developed a **model** to help us understand the structure of matter. This model is called the **particle theory of matter**. This theory states that matter is made up of particles that are in constant motion. The more energy the particles have, the faster they move and the further apart they are. Because the movement of the particles is related to their **kinetic energy** (the energy due to movement), this theory is also known as the kinetic theory of matter or the kinetic particle theory.

particle theory of matter
a theory that states that all matter is made up of particles that are in constant motion

kinetic energy
the energy of an object due to its motion

☆ ACTIVITY

The particle model simulation

Go to the PhET website and select the 'States of matter: Basics' simulation. Then click on 'States'.

1 Choose a type of atom or molecule.
2 Change the temperature to °C.
3 Press play.
4 Observe the movement of the particles.
5 Add heat and observe how the movement of the particles changes.
6 Remove heat by cooling and observe how the movement of the particles changes.
7 What is the lowest temperature that can be reached?

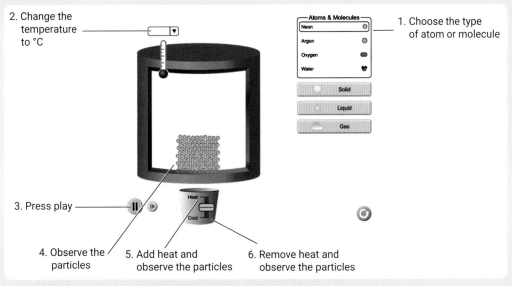

▲ FIGURE 3.2.3 The particle model simulation

Weblink
PhET: States of matter

3.2 LEARNING CHECK

1 **State** the particle theory of matter.
2 **Draw** a labelled diagram to show the structure of water in a beaker.
3 **Suggest** why other scientists might not have initially believed Democritus's explanation of matter being made up of particles.
4 An analogy is something that can be used to explain something else. One analogy of the structure of matter is students in school. **Explain** how this analogy models the structure of matter.

3.3 States of matter

BY THE END OF THIS MODULE, YOU WILL BE ABLE TO:
✓ describe the structure of solids, liquids and gases by applying the particle theory of matter.

Video activity
Solids, liquids and gases

GET THINKING

Prepare a concept map, similar to the one in Figure 3.3.1, to summarise the content in this module. Add to your concept map as you read the information. If you wish, you can include diagrams to help your understanding.

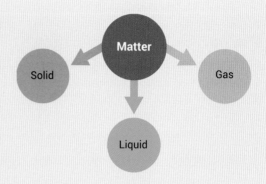

▲ FIGURE 3.3.1 A concept map for the states of matter. Add to it as you read through the module.

States of matter

state of matter
one of the forms in which matter can exist – solid, liquid or gas

Although all matter is made up of particles, there is a huge variety in the types of matter. One way of classifying matter is by the **state of matter** – whether it is a solid, a liquid or a gas.

▲ FIGURE 3.3.2 Matter can exist as solids, liquids or gases.

You are already familiar with solids, liquids and gases. But what makes a solid a solid, a liquid a liquid, or a gas a gas? The particle theory of matter is useful in answering this question because it allows us to understand the structure of the different states of matter.

Solids

The particles in a **solid** are held close to one another in a fixed arrangement. This is because the strong **forces of attraction** between the particles pull them together. In a solid, the particles only have a small amount of energy, which allows the particles to vibrate where they are. However, they do not have enough energy to move away from one another.

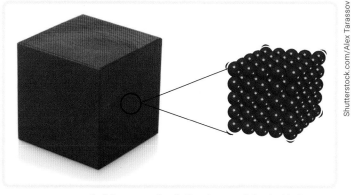

▲ FIGURE 3.3.3 Solids are made of vibrating particles held close to one another.

solid
a state of matter in which the particles vibrate in fixed positions close to each other

force of attraction
a force that pulls objects together

liquid
a state of matter in which the particles are close together but unable to break free of each other

Liquids

The particles in a **liquid** have more energy than those in a solid. Although the particles are still attracted to one another, they have enough energy to move faster and overcome the attraction. Because of this, the particles in a liquid are not held in a fixed place. Instead, they can move past one another as they break and then re-form forces of attraction with other particles.

▲ FIGURE 3.3.4 The particles in a liquid can move past one another.

Gases

Gases may be the hardest state of matter to understand because, in many instances, we cannot actually see them. This is because the particles in gases are as far apart as possible.

gas
a state of matter in which the particles are very far apart and move with a lot of energy

The particles in a gas have even more energy than those in liquids. This means that they are moving faster and can break free from one another. In fact, they keep moving until they collide with something else, either another particle or the wall of the container. They then bounce off the object and move in another direction. Therefore, the particles of a gas are spread throughout the container that the gas is held in.

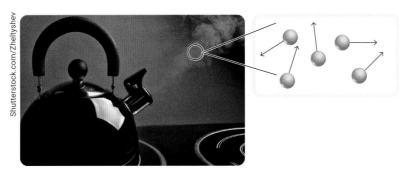

▲ FIGURE 3.3.5   The particles in a gas are moving fast and are as far apart as possible.

ACTIVITY

Modelling the states of matter

Choose a method from A–D to teach other students about the structure of solids, liquids and gases. You can use the following materials depending on your chosen resource:

- 3 or 4 ice cubes on a shallow dish such as a Petri dish
- water in a beaker
- water in a kettle
- white paper
- green paper
- magnifying glass
- tablet or phone to take photos and/or videos
- materials to make models.

As you complete the next modules, you may add more information to your resource.

A Using a green screen

1. Use a program such as PowerPoint or Keynote to create animations that model the structures of solids, liquids and gases.
2. Stick a green piece of paper over the lens of a magnifying glass.
3. Video the magnifying glass moving over the ice cube, water and then steam from the kettle.
4. Link your animations with your green screen videos so that the magnifying glass 'sees' the structure of the solid, liquid or gas.
5. Add a voice-over to describe the structure.

B Making a model

1. Create a model that represents the structures of solids, liquids and gases. For example, you could use marbles to represent the particles and find a way to make them move in the manner that the particles would.

C Making a YouTube video

1 Create a short YouTube video explaining the structure of solids, liquids and gases. What format will you use? For example, you could use cartoon characters, screen recording, real characters explaining or stop-motion videos.

D Role play

1 Use other students to act as particles in matter.
2 Be the director of the play to tell the actors how to move and behave when modelling a solid, a liquid or a gas.

Summarising states of matter

▼ TABLE 3.3.1 Summarising the structure of solids, liquids and gases

Characteristic	State of matter		
	Solid	**Liquid**	**Gas**
Energy	Low	Medium	High
Forces of attraction	Strong	Medium strength	Very weak
Proximity to other particles	As close together as possible	Relatively close together	As far apart as possible
Arrangement of particles	Orderly arrangement in a fixed position	Irregular arrangement	Spread throughout the container
Movement of particles	Vibrating on the spot	Sliding past one another	Moving quickly in a straight line until they hit another particle or wall, then they rebound and continue moving

3.3 LEARNING CHECK

1 **List** the three states of matter.
2 **Identify** each of the diagrams in Figure 3.3.6 as either solid, liquid or gas based on the arrangement of the particles.

a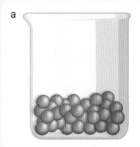
b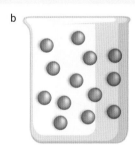
c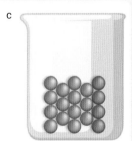

▲ FIGURE 3.3.6

3 **Construct** a Venn diagram to compare the structure of solids, liquids and gases.
4 Some solids, such as your desk, are very hard and rigid. Others, such as your jumper, are softer and more flexible. **Discuss** the possible reasons for these differences based on the particle theory of matter.

3.4 Properties of solids

BY THE END OF THIS MODULE, YOU WILL BE ABLE TO:
- ✓ list the main properties of solids
- ✓ explain the properties of solids.

> **GET THINKING**
>
> Skim read this module, paying particular focus to the headings. Can you list the properties of solids? Can you predict why solids have these properties? Now read the module in detail – were you correct?

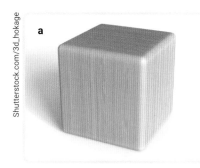

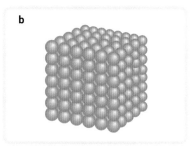

▲ **FIGURE 3.4.1** (a) A wooden cube has a fixed shape. (b) The shape of the cube is due to the arrangement of the particles

Properties of solids

Although there are many different solids, they all have the same key properties.

- Solids have a fixed shape.
- Solids have a fixed volume.
- Solids do not flow.
- Solids cannot be compressed.

We can use the particle theory of matter, and hence the structure of solids, to explain these properties.

Solids have a fixed shape

If you put a piece of cake on your plate, it will remain in that shape (until you eat it!). In the same way, a book stays the same shape, as does a wooden picture frame. This is what we mean by having a fixed shape.

The reason that solids have a fixed shape is that their particles are held together by strong forces of attraction. This keeps the particles close together and prevents them from moving into different positions (Figure 3.4.1).

You might be thinking 'But an ice cube will change shape when it melts' or 'A book changes shape when I turn a page'. You are correct. However, a solid doesn't change its shape without something else acting on it.

Solids have a fixed volume

volume
the amount of space occupied, measured in litres (L) or millilitres (mL)

The **volume** of something is the amount of space that it occupies. A solid occupies a set amount of space because the forces of attraction between the particles hold them the same distance from one another.

One way to measure volume is to measure the sides and use this to calculate the volume. This works for regular shapes; however, many solids are not regular shapes. An alternative method is to put the solid into water in a measuring cylinder. The volume of the solid can be calculated from the volume that the water increases by (Figure 3.4.2).

Solids do not flow

When something **flows**, it moves from one place to another in a steady stream. To do this, the particles must be able to move past one another into a new place. Solids are unable to flow because the forces of attraction mean that the particles are held together and can't move into a new place.

flow
move from one place to another in a steady stream

Solids cannot be compressed

If an object is **compressed**, it is squashed so that it takes up less space. The ability to do this is defined as **compressibility**. In a solid, the particles are already held closely together and cannot be easily pushed closer together. Therefore, solids can't easily be compressed.

You can try this yourself by trying to squash a book or a piece of wood. You may end up breaking it, but you can't squash it. Some objects, such as sponges and tennis balls, are deceiving and it may appear that you can compress them. In fact, these objects have air inside them too, and when you squash the object, you are compressing or squeezing out the air. The solid part of the object has not been compressed.

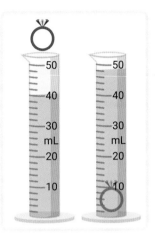

▲ **FIGURE 3.4.2** The volume of a solid object can be determined from the volume of water it displaces. The water level rose from 40 mL to 48 mL, so the ring has a volume of 8 mL.

compress
to squash something so it takes up less space

compressibility
the ability to be compressed (or squashed)

▲ **FIGURE 3.4.3** Luckily, solid bricks do not compress. Otherwise, our houses would get shorter as the bricks become squashed!

Quiz
What are the properties of solids?

3.4 LEARNING CHECK

1. **List** the main properties of solids.
2. **Explain** why an ice cube is unable to flow.
3. **Apply** your understanding of the particle theory to explain why an apple has a fixed shape and volume (until it is eaten).
4. A metal cube is harder than a piece of wood. **Compare** the strength of the forces of attraction between the particles to suggest a reason for the difference in hardness.
5. The chair that you are sitting on is made of different solids. **Identify** and **explain** the properties of solids that make them suitable for use as a chair.
6. Brie cheese is very soft and is easy to cut and spread. Would you classify brie cheese as a solid? **Justify** your answer.

3.5 Properties of liquids

BY THE END OF THIS MODULE, YOU WILL BE ABLE TO:
- ✓ list the main properties of liquids
- ✓ explain the properties of liquids.

Interactive resource
Drag and drop: Solids v liquids

Extra science investigation
Particle movement in liquids

GET THINKING

Water is a liquid that you are very familiar with. You use it for washing, drinking and even swimming. What properties does water have that make it a liquid?

Properties of liquids

For a substance to be classified as a liquid, it must have the following properties.
- Liquids take the shape of the container.
- Liquids have a fixed volume.
- Liquids can flow.
- Liquids cannot be easily compressed.

Just like with solids, the way the particles are arranged and behave in liquids determines these properties.

▲ FIGURE 3.5.1 Liquids spread out to take the shape of the container.

Liquids take the shape of the container

When a liquid is poured into a container, it spreads out to the edges of the container (Figure 3.5.1). This is because the particles can move over one another, filling the spaces until all the shape is filled.

Liquids have a fixed volume

The particles in a liquid are attracted to one another by forces of attraction. These forces keep the particles close to one another, so the volume of a liquid remains constant.

This property is easy to understand when we look at a sample of liquid in one container. However, it also applies when we move that same liquid into another container. For example, you use a measuring cup to measure 250 mL of milk when baking a cake. When you pour that milk into the mixing bowl, it will spread out in the larger container (Figure 3.5.2). Yet there is still only 250 mL of milk.

▲ FIGURE 3.5.2 Even when the liquid is poured into a different container, it still has the same volume.

▲ FIGURE 3.5.3 Liquids can flow, allowing us to pour a glass of soft drink.

Liquids can flow

The energy of the particles in a liquid allows them to overcome the forces of attraction enough to move past one another. As they move, they can occupy a new space, taking the liquid from one place to another. This movement is called flowing.

Liquids cannot be easily compressed

Just like solids, liquids are made up of particles that are held closely together. There is very little space between them, which means that the particles can only be pushed a little bit closer together. Therefore, liquids can only be slightly compressed, or squashed, and this requires a large amount of force (Figure 3.5.4).

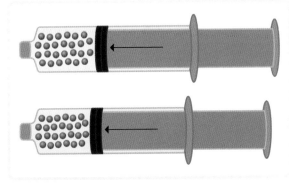

▲ FIGURE 3.5.4 Liquids cannot be easily compressed because the particles are already close together.

3.5 LEARNING CHECK

1. **List** the main properties of liquids.
2. **Explain** why liquids can flow.
3. Some people prefer to sleep on a waterbed, which is a mattress filled with water rather than a solid material. Use the properties of liquids to **explain** why waterbeds may be more comfortable than traditional mattresses.
4. **Draw** a labelled diagram to show what is happening at a particle level when water is poured into a glass.
5. **Describe**, from the point of view of a particle, how water travels from a water tank, through a hose and onto the ground while watering a garden.

Chapter 3 | Matter: solids, liquids and gases

3.6 Properties of gases

BY THE END OF THIS MODULE, YOU WILL BE ABLE TO:
- ✓ list the main properties of gases
- ✓ explain the properties of gases.

GET THINKING

Long thin balloons filled with a gas such as air can be used to make different shapes, such as the animals in Figure 3.6.1. What properties do gases have that makes these creations possible? Discuss your ideas with a partner – how many properties can you list?

Video activity Properties of gases

▲ FIGURE 3.6.1 Balloon animals

Properties of gases

Remember that the particles of gases have more energy than particles in liquids or solids, meaning they are moving faster and are as far apart as possible. This is significantly different from the structure of liquids and solids, resulting in different properties.

Gases all have the same key properties.
- Gases take the shape of the container.
- Gases take the volume of the container.
- Gases can flow.
- Gases can be compressed.
- Gases can diffuse.

▲ FIGURE 3.6.2 The particles in a gas are spread out as far as possible, filling the container.

Gases take the shape and volume of the container

Gas particles continue to move until they collide with, and rebound from, the walls of the container. Therefore, gas particles will fill the container, taking its shape and volume. This may be hard to picture because we cannot see most gases. However, it is easy to understand with the example of a balloon, where the gas particles spread out and keep the balloon pushed out into its shape (Figure 3.6.2).

Gases can flow

When a substance flows, it moves from one place to another. Gases can flow because their particles can move into new spaces because of their motion and weak forces of attraction.

When a gas is in a container, such as LPG in a gas bottle, the gas particles are held in the container. When the container is opened, rather than hitting a wall in that area, the particles continue moving out of the container. More and more gas particles move through the opening, allowing the gas to flow out of the container.

Gases can be compressed

Unlike solids and liquids, gases have a large amount of space between their particles due to the particles being spread as far apart as possible. This means that gas particles can be pushed closer together, compressing the gas (Figure 3.6.3).

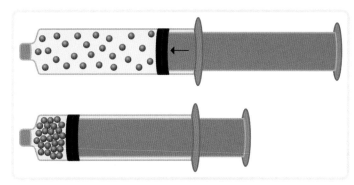

▲ **FIGURE 3.6.3** When gases are compressed, the particles are pushed closer together.

diffusion
the movement of particles from an area of high concentration to an area of low concentration

Gases can diffuse

If someone is cooking dinner, you can often smell it as soon as you walk into the room, even though the food is still on the stove. This is due to **diffusion**. Diffusion is the movement of particles from an area of high concentration to an area of low concentration until they are evenly spread out.

Diffusion occurs in gases because the particles can move freely from one area to another. This means that gas particles, such as those producing the delicious smell of dinner, will spread out across the room (Figure 3.6.4).

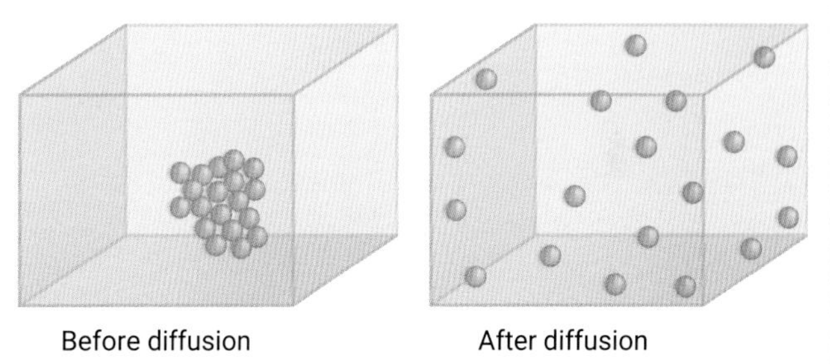

▲ **FIGURE 3.6.4** Diffusion of gases involves gas particles moving until they are evenly spread out.

3.6 LEARNING CHECK

1. **List** the main properties of gases.
2. **Explain** why gases take the shape of their container.
3. When the plunger of a sealed syringe, such as the one in Figure 3.6.5, is depressed, the gas is compressed. **Explain** why it gets difficult to compress the gas when the volume becomes very small.

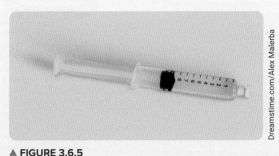

▲ **FIGURE 3.6.5**

4. Natural gas does not have an odour. Gas companies that supply natural gas to households add the harmless chemical mercaptan. Mercaptan smells like rotten eggs to make it easy to detect gas leaks. **Explain** how mercaptan can achieve this function and possibly save lives.
5. **Draw** a labelled diagram to explain why a cricket ball cannot be squashed, but a tennis ball can.
6. **Create** a Venn diagram to compare the properties of solids, liquids and gases.

3.7 Changing state

BY THE END OF THIS MODULE, YOU WILL BE ABLE TO:
- ✓ name the changes in state
- ✓ explain why changes of state occur.

Video activity
Changing states of matter

Interactive resources
Simulation: Phase changes
Crossword: Changing states

Extra science investigation
Changes of state and the weather

GET THINKING

Can you think of an example where a liquid becomes a solid or a gas? What causes it to change? What is happening to the particles during the change? As you learn about changes of state in this module, reflect on your answers. Were you correct? If not, what stopped you from correctly understanding the process?

Changing state

If you leave an ice cube in sunlight, it will become a liquid and eventually all of it will 'disappear' as it forms a gas. This is evidence that matter can change its state. However, for this to happen, energy needs to be either added or removed.

▲ **FIGURE 3.7.1** An ice cube can change from a solid to a liquid and then to a gas.

Heating a substance

When a substance is heated (e.g. by putting it in the oven), the particles gain energy and move faster and further away from the other particles. This allows the particles to change their arrangement from the fixed arrangement in a solid, to moving over one another in a liquid, to moving quickly away from one another in a gas.

The process of a solid absorbing energy and becoming a liquid is called **melting**. Each substance requires a particular amount of energy to melt, and the temperature at which enough energy is provided is called the **melting point**. The process of changing into a liquid is called **liquefication**.

melting
the process of changing from a solid to a liquid

melting point
the temperature at which a substance changes from a solid to a liquid

liquefication
the process of changing into a liquid

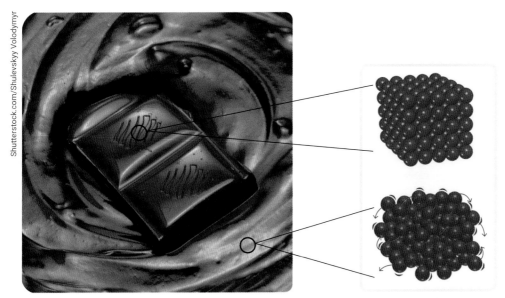

▲ **FIGURE 3.7.2** Solid chocolate melts into liquid chocolate when heated. This gives the particles more energy to overcome the forces of attraction, move further apart and flow.

At any temperature, some of the particles at the surface of a liquid have enough energy to move away from the other particles and become a gas. This process is called **evaporation**.

If the liquid absorbs enough energy, it will reach the **boiling point**, the temperature when all of the liquid becomes a gas. At this temperature, the particles have enough energy to overcome the forces of attraction and move away from the other particles. This means that it has formed a gas. This process is called **boiling**.

Some substances, such as carbon dioxide, do not form liquids under normal conditions. Instead, their particles move from an orderly arrangement in the solid to being spread out in a gas. This process is called **sublimation**.

Boiling, evaporation and sublimation are all examples of **vaporisation** because they form vapours, or gases.

evaporation
the process of changing from a liquid to a gas at a temperature lower than the boiling point

boiling point
the temperature at which all of a substance changes from a liquid to a gas

boiling
the process of changing from a liquid to a gas at the boiling point

sublimation
the process of changing from a solid directly to a gas

vaporisation
the process of forming a gas; evaporation, boiling or sublimation

▼ **TABLE 3.7.1** The melting and boiling points of some common substances

Substance	Melting point (°C)	Boiling point (°C)
Dry ice (frozen carbon dioxide)	−57	−79 (sublimes)
Water	0	100
Table salt	804	1413
Gold	1064	2970
Diamond	3550	

Chapter 3 | Matter: solids, liquids and gases

Cooling a substance

As heat is removed from a substance (e.g. by putting it in the refrigerator), the particles lose energy. Therefore, they move slower and no longer have the energy to remain apart from one another. This means that a gas will become a liquid (**condensation**) or a liquid will become a solid (**solidification** or **freezing**). Substances that sublime when heated (change from a solid to a gas) will change from a gas directly back to a solid. This process is called **deposition**.

▲ **FIGURE 3.7.3** Icicles form when liquid water freezes, producing a solid.

condensation
the process of changing from a gas to a liquid

solidification
the process of changing from a liquid to a solid

freezing
the process of changing from a liquid to a solid

deposition
the process of changing from a gas to a solid

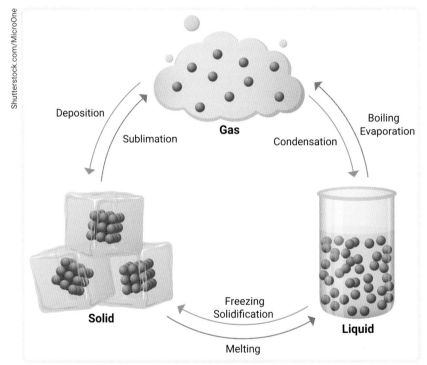

▲ **FIGURE 3.7.4** A summary of the changes of state

Using changes of state

We can apply our understanding of changes of state, and how they occur, in many ways.

When we wash our clothes, they become wet with water, a liquid. For the clothes to become dry, the liquid water needs to change to a gas. One way to achieve this is by putting them on the clothesline in sunlight. The Sun's energy heats the particles, giving them enough energy to become a gas. Another method is to put the clothes in the dryer, which uses electricity to warm the clothes and evaporate the water.

Another example is the production of ice-cream. Ice-cream is a mixture of milk, cream, sugar and flavouring. These are mixed, making a liquid that can be solidified to become ice-cream. Any method of cooling the mixture enough will produce ice-cream. One method is to pour liquid nitrogen, which has a temperature of −196°C, over the liquid. The very cold liquid nitrogen freezes the ice-cream, and then evaporates, leaving only the ice-cream behind (Figure 3.7.6). The very low temperature of the liquid nitrogen makes it dangerous, so it should only be used by adults wearing protective glasses, gloves and clothing.

▲ FIGURE 3.7.5 The Sun's energy can be used to change the liquid in wet fabric to a gas.

▲ FIGURE 3.7.6 The coldness of liquid nitrogen can be used to freeze ice-cream.

3.7 LEARNING CHECK

1 **Match** each term with its definition.

Term	Definition
Melting	• The process of changing from a gas to a liquid
Condensation	• The process of changing from a gas to a solid
Liquefication	• The process of changing into a liquid
Deposition	• The process of changing from a liquid to a solid
Evaporation	• The process of forming a solid
Freezing	• The process of forming a gas; evaporation, boiling or sublimation
Boiling	• The process of changing from a solid to a liquid
Sublimation	• The process of changing from a liquid to a gas at a temperature lower than the boiling point
Vaporisation	• The process of changing from a solid to a gas
Solidification	• The process of changing from a liquid to a gas at the boiling point

2 **List** the changes of state that require the removal of heat.
3 **Explain** why water vapour will condense on a car window on a cold morning.
4 When clothes dry on the clothesline, is the change of state evaporation or boiling? **Justify** your answer.
5 **Explain** why steam will burn you more severely than boiling water.
6 **Draw** a flow chart to represent the changes of state.
7 Absolute zero (−273°C) is the lowest temperature that is possible. **Suggest** what happens to particles at this temperature and why it is not possible to get even lower temperatures.
8 The melting point of butane (a highly flammable gas) is −138°C, whereas the melting point of oxygen is −219°C. Provide a possible reason for the difference between these melting points.

3.8 Collecting drinking water

SCIENCE AS A HUMAN ENDEAVOUR

BY THE END OF THIS MODULE, YOU WILL BE ABLE TO:
- ✓ explain how drinking water can be collected from the air
- ✓ explain how drinking water is produced on a boat.

Collecting drinking water in the desert

Water is a necessity of life – without it we could not survive! But it is not only human life that needs water – plants and animals also need access to water. However, many regions around the world are suffering drought. Farmers are having to transport water in to keep animals alive, and their crops cannot survive. Other areas do not even have access to safe drinking water, forcing millions of people to drink contaminated water. According to the World Health Organization, in 2019, at least 2 billion people were forced to use drinking water that was contaminated with faeces. Also, by 2025, half of the world's population will be living in **water-stressed** areas.

Scientists are constantly testing new methods of accessing water and producing water that is safe to drink. One area of research is using **water vapour** in the air of the desert. Scientists started by studying the plants and animals that survived in the desert so that they could learn about the methods used to obtain water. This knowledge was used by the scientists to develop new technologies and devices that copied the process. The devices work by collecting the water vapour from the air at night when the **humidity** is greatest. It is then cooled so that the water vapour condenses, forming liquid water.

water stress
when the amount of water needed is greater than the amount of water available

water vapour
the gaseous form of water

humidity
the amount of water vapour in the air

▲ FIGURE 3.8.1 Getting water from the desert air is becoming a reality.

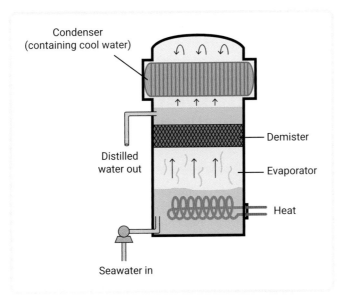

▲ **FIGURE 3.8.2** A simple water evaporator used to produce drinking water on ships

Producing drinking water on a boat

Having a reliable supply of drinking water can also be a problem if you are travelling on a boat. This problem is easy to overlook because you are surrounded by water. However, the seawater is not suitable for drinking, and it may not be feasible to carry enough bottled water to last the duration of the trip.

One option for producing drinking water mimics the water cycle, in which water evaporates into the atmosphere, forms clouds, and then condenses to return to the ground as rain. On ships, a water evaporator achieves the same steps. A heat source, such as the engine, warms seawater, causing it to evaporate. The water vapour is then cooled, allowing it to condense back into liquid water. As only the water evaporates, the salt in the seawater is left behind and so the water that is collected is suitable for drinking.

3.8 LEARNING CHECK

1. What state of matter is water vapour?
2. **Explain** why the water vapour is cooled after it is collected.
3. Do you think the humidity will be higher in a desert or in a rain forest? **Explain** your answer.
4. **Discuss** whether you think it is worthwhile trying to collect water from the desert air. In your discussion, consider both the advantages and the disadvantages.
5. Use a flow chart to model the states of matter and changes of state that occur when an evaporator is used to produce drinking water on a ship.
6. **Predict** a disadvantage of using an evaporator to produce drinking water on a ship.
7. **Compare** the methods of producing drinking water in the desert and on a ship.

SCIENCE INVESTIGATIONS **3.9** # Organising and representing data

SCIENCE SKILLS IN FOCUS

IN THIS MODULE, YOU WILL FOCUS ON LEARNING AND IMPROVING THESE SKILLS

- organising data in an effective table
- representing data with a graph.

Tables and graphs are important tools for organising and visualising data. In Chapter 2, you looked at how to construct tables and graphs. In this module, you will have the opportunity to practise these skills.

- When constructing a table:
1 include an informative title
2 record your data in the columns. Draw the lines with a ruler if the table is hand drawn
3 ensure that the headings on the columns describe the data
4 state the units for the data in the column headings only
5 put the independent variable (the one you changed) in the first column
6 put the dependent variable (the one you measured or found out) in the other columns.

- When constructing a graph:
1 include an informative title
2 draw clear horizontal and vertical axes
3 include labels on the axes describing the data
4 state the units for the data in brackets with the axis labels
5 plot the independent variable (the one you changed) on the *x*-axis (horizontal)
6 plot the dependent variable (the one you measured or found out) on the *y*-axis (vertical)
7 if the data are numerical, increase the scale in even increments
8 in most cases, only graph the average results
9 if you have more than one set of data on a graph, use a key to clearly identify each dataset
10 for line graphs, use a line of best fit/smooth curve
11 for column graphs, have a space between each column
12 make sure the graph uses as much of the grid as possible.

INVESTIGATION 1 : MELTING AND BOILING POINTS OF WATER

AIM

To determine the melting and boiling points of water

YOU NEED

- ☑ crushed ice
- ☑ 250 mL beaker
- ☑ tripod
- ☑ gauze mat
- ☑ heatproof mat
- ☑ Bunsen burner
- ☑ thermometer or temperature probe
- ☑ retort stand, boss head and clamp
- ☑ stirring rod

WHAT TO DO

1 Set up the equipment as shown in Figure 3.9.1.
2 Draw a table to record the temperature at different times. Remember to put the independent variable (time) in the first column and the dependent variable (temperature) in the second column. Don't forget a title and units in the column headings.
3 Half-fill the beaker with crushed ice. Add enough water to just cover the ice. Stir well with the stirring rod.
4 Place the beaker on the gauze mat and heat the beaker using a blue flame on the Bunsen burner.
5 Record the temperature every minute if you are using a thermometer, or every 15 seconds if you are using a temperature probe.
6 Continue recording the temperature until the water has been boiling for five consecutive minutes.
7 Turn the Bunsen burner off and keep recording the temperature of the water every minute or 15 seconds for 10 minutes.

Video
Science skills in a minute: Organising data

Science skills resource
Science skills in practice: Organising and representing data

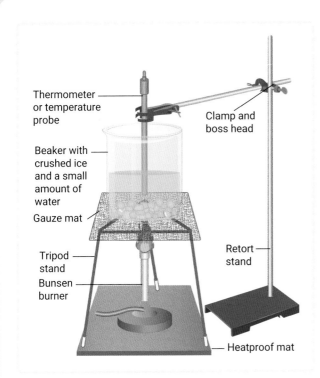

▲ FIGURE 3.9.1   How to set up your equipment

 Broken glass can cut skin. Clean up any broken glass immediately and put it in the glass bin.

Hot substances and objects can cause burns. Do not touch hot objects. Use equipment such as tongs to move hot equipment.

Wear safety glasses and lab coats at all times.

WHAT DO YOU THINK?

1 **Explain**, using the terms 'particles' and 'energy', why the ice melted.

2 **Explain**, using the terms 'particles' and 'energy', why the water boiled.

3 **Draw** a graph of the temperature at each time you measured it.
   a What type of graph should you use? How do you know what type of graph to use?
   b Give your graph a title: it may be the same as your table's title.
   c Label the axes, show units and use an even scale.
   d Draw a line of best fit.

4 What does the graph tell you?
   a What was the temperature when the ice was melting?
   b **Describe** what happened to the temperature of the water as the ice melted.
   c When did the temperature start to rise? What happened first?
   d What was the temperature when the water was boiling?
   e **Describe** what happened to the temperature as the water boiled.

5 **Describe** what is happening to the particles in the water in terms of energy and movement when the:
   a water is solid.
   b ice is melting.
   c water is all liquid and being heated.
   d water is boiling.
   e water is a gas.

6 What can you **conclude** from your experiment? Write a sentence to summarise what you learned about the melting and boiling points of water.

INVESTIGATION 2: MELTING POINTS OF MATTER

AIM

To determine the melting point of six materials, and the boiling point of water

YOU NEED

- wax block
- gallium block
- crayon
- ice cube
- piece of chocolate
- marshmallow
- 6 beakers
- hot plate
- infrared thermometer

Chapter 3 | Matter: solids, liquids and gases

WHAT TO DO

1. Make a prediction about the temperature at which each material will melt (the melting point). Record this in the results table.
2. Place the wax block, crayon, ice cube, piece of chocolate and marshmallow into six separate beakers.
3. Place the beakers onto the hot plate, but do not turn the hot plate on.
4. Wait 3–4 minutes to see if any of the materials begin to melt at room temperature. Record your observations in the 'Actual melting point' column in the results table.
5. Turn on the hot plate and move the dial to the lowest setting.
6. Begin measuring the temperature of the glass beakers using the infrared thermometer. Measure the temperature at frequent intervals (e.g. 20 seconds).
7. Carefully observe whether the materials in the beakers are beginning to melt. As soon as you see part of the material has melted, record the temperature in the results table.
8. Continue to monitor the temperature and materials while slowly adjusting the setting of the hot plate to make it warmer.
9. Once all six materials have completely melted, remove the beakers containing the wax, crayon, chocolate and marshmallow. Leave the melted ice on the hot plate.
10. Increase the temperature of your hot plate further until the water reaches boiling point. Record the temperature at which this occurs.

RESULTS

Copy and complete the results table.

Material	Predicted melting point (°C)	Actual melting point (°C)	Boiling point (°C)
Wax			
Crayon			
Chocolate			
Marshmallow			
Ice cube			

WHAT DO YOU THINK?

1. List the materials in order of lowest melting point to highest melting point. Was this what you expected?
2. Did any of the materials melt at room temperature?
3. What was the state of the materials at the beginning of the experiment? What was their state at the end of the experiment?
4. Why did a change in state occur?
5. How do your results compare to those of the group next to you? Did your materials melt at the same temperature?

CONCLUSION

Compare the melting points of all six materials, discussing whether any were similar.

3 REVIEW

REMEMBERING

1. **Describe** the arrangement of particles in a:
   a solid.
   b liquid.
   c gas.

2. Copy and **complete** the following table by placing a tick for each property that solids, liquids and gases exhibit.

Property	Solids	Liquids	Gases
Have a fixed shape			
Have a fixed volume			
Can be compressed			
Take up space			
Have mass			
Able to flow			

3. What term describes the following situations?
   a A solid turns into a liquid.
   b A gas turns into a liquid.
   c A liquid turns into a solid.
   d A gas turns into a solid.

4. What state of matter (solid, liquid or gas) is formed in the following situations?
   a A gas condenses.
   b A liquid evaporates.
   c A solid sublimes.

5. **Identify** the following sentences as true or false. If they are false, rewrite them to make them true.
   a Liquids keep the same volume, but solids don't.
   b Gases take up the same amount of space as liquids.
   c Liquids and solids cannot be compressed, but gases can.
   d Solids maintain their shape, whereas liquids and gases only maintain their volume.
   e A set quantity of matter will have the same mass whether it is solid, liquid or gas.
   f There is much more space between the particles in a gas than there is between the particles in a solid.
   g Gases are always made of much smaller particles than solids are.

UNDERSTANDING

6. Substance X is in a container. When the container is tipped, substance X flows. What else would you need to know to determine what state substance X is in?

7. **Identify** whether the following state changes require the addition or removal of energy.
   a Boiling
   b Condensing
   c Freezing
   d Evaporating

8. **Explain** the difference between evaporation and boiling.

9. For each of the following pairs of options, which option would have particles with the most energy? **Justify** your answer.
   a Solid butter or melted butter
   b Honey or the smell of honey

10. **Explain** how the particles in ice move compared with the particles in water and steam.

APPLYING

11. **Apply** the properties of each substance to identify them as solids, liquids or gases.
    a A rock
    b Fog
    c Honey
    d Fur
    e An iceberg
    f The smell of a flower
    g Butter

12. **Explain** why you can walk through a sheet of water (such as a waterfall) but not through a sheet of glass.

13 **Explain** why bread that is baking smells so much better than cold bread.

14 When water boils there are bubbles in it. What do you think is in the bubbles? **Explain** your answer.

15 In each of the following situations, which process requires the addition of the most heat?
   a Melting ice or melting the same amount of gold if they are starting from the same temperature
   b Boiling vegetable oil or boiling water
   c Subliming carbon dioxide or subliming iodine (Hint: What other information do you need to be able to answer this?)

16 'Gases exert much more pressure than solids.' Decide whether this statement is true or false and **explain** your answer.

17 **Explain** why cordial diffuses in water quickly, but if it is added to solid ice, it doesn't mix.

EVALUATING

18 Why does the bathroom mirror fog up after you have a hot shower without the fan on?

19 **Suggest** why dogs pant when they are hot.

20 As you go up into the atmosphere, such as up a mountain or in an aeroplane, the air gets thinner and the pressure decreases. This is why your ears 'pop' – they need to equalise the pressure inside and outside.
   a **Compare** the arrangement of particles in the air at the top of a mountain and at the bottom of the mountain.
   b **Explain** why you are not allowed to take a pressurised container of gas on an aeroplane.

21 Why is it easier to float in a swimming pool while holding your breath than when you let it out?

CREATING

22 **Create** a concept map of all the key terms in this unit. Make as many links as possible between them.

BIG SCIENCE CHALLENGE PROJECT #3

1 Connect what you've learned

In this chapter, you have learned about the states of matter, and how matter changes state.

Brainstorm what you remember by making a list of the key terms.

2 Check your thinking

Explain, in terms of particle theory and states of matter, what is happening when someone sweats. Think about whether this process absorbs heat or releases heat. Use this to explain how sweating can cool someone down.

3 Make an action plan

Conduct some research into anhidrosis in humans or horses. Think about what you have learned in this chapter so that you can understand why the condition is life-threatening. Learn about how the condition is managed, and alternative methods that are used to cool down people with anhidrosis.

4 Communicate

Use your knowledge and understanding to create a brochure about anhidrosis. Your brochure should be suitable to have in a doctor's (or vet's) waiting room and must be clear and easy to understand.

Pure substances and mixtures

4.1 Classifying matter (p. 98)
Matter can be classified as a pure substance, or a mixture, based on the type of particles.

4.2 Pure substances: going further (p. 102)
Pure substances can be elements or compounds.

4.3 Solubility (p. 104)
The solubility of a substance describes how much can dissolve in a certain volume.

4.4 Solutions (p. 106)
A solution is produced when a solute dissolves in a solvent.

4.5 Suspensions (p. 110)
A suspension is produced when an insoluble solute is too heavy to remain dispersed throughout the solvent.

4.6 Colloids (p. 112)
Colloids are produced when an insoluble solute is light enough to remain dispersed throughout the solvent.

4.7 SCIENCE AS A HUMAN ENDEAVOUR: Air quality (p. 114)
Air is a mixture that is monitored for the amount of pollutants is contains.

4.8 SCIENCE INVESTIGATIONS: Using observations to make classifications (p. 116)
Observing mixtures

BIG SCIENCE CHALLENGE #4

▲ FIGURE 4.0.1 Fish need to obtain oxygen from water.

You probably know that we need to breathe in air to give our bodies oxygen. But what about fish – do they breathe? Fish also need oxygen, but they need to get it from the water!

▶ Why is it important that oxygen dissolves in water?

▶ What factors must be monitored to ensure that there is enough oxygen in the water for fish in an aquarium?

▶ How would life on Earth be different if oxygen couldn't dissolve in water?

#4 SCIENCE CHALLENGE ACCEPTED!

At the end of this chapter, you can complete the Big Science Challenge Project #4. You can use the information you learn in this chapter to complete the project.

Assessments
- Prior knowledge quiz
- Chapter review questions
- End-of-chapter test
- Portfolio assessment task: Research project

Videos
- Science skills in a minute: Observations (4.8)
- Video activities: Elements and compounds (4.2); Solutions (4.4); Measuring air quality (4.7)

Science skills resources
- Science skills in practice: Observing to make classifications (4.8)
- Extra science investigations: Design your own solubility investigation (4.3); How much sugar? (4.4)

Interactive resources
- Simulations: Build a compound (4.2); Concentration of solutions (4.4)
- Drag and drop: Classifying mixtures and pure substances (4.1); Solubility (4.3)
- Crossword: Solutions, suspensions and colloids (4.6)
- Label: Compound, element or mixture? (4.2)

Nelson MindTap

To access these resources and many more, visit:
cengage.com.au/nelsonmindtap

4.1 Classifying matter

BY THE END OF THIS MODULE, YOU WILL BE ABLE TO:
- ✓ define matter, pure substance and mixture
- ✓ use diagrams to model pure substances and mixtures
- ✓ give examples of pure substances and mixtures
- ✓ classify a substance as a pure substance or a mixture
- ✓ compare pure substances and mixtures.

GET THINKING

A glass of seawater looks the same as a glass of tap water – it is clear and colourless. Yet, after swimming in the ocean, it often feels as though there is something scratching your skin. Discuss these observations with a partner – what inferences can you make?

▲ **FIGURE 4.1.1** After swimming in the ocean, it can feel as though something is scratching your skin.

What are pure substances and mixtures?

In Chapter 3, you learned that matter is anything that has a mass and occupies space. In fact, everything around you is made up of matter. This includes the air, water, your body, the road, a pen, the food you eat and even the screen of your phone. While it is useful to have a name for all these substances, matter is such a broad group that it doesn't tell you much about the properties of them. Therefore, scientists have different ways of classifying matter based on common properties. One method of classifying substances is according to whether they are **pure substances** or **mixtures**.

We classify substances as pure substances or mixtures based on the type of particles that they are made up of. In a pure substance, all the particles are the same. For example, in a glass of pure water, there are only water particles present. However, mixtures are made up of different types of particles. Seawater is a mixture because it is made up of water particles as well as salt particles.

pure substance
a substance made up of the same type of particle

mixture
a substance made up of different types of particles that are physically combined

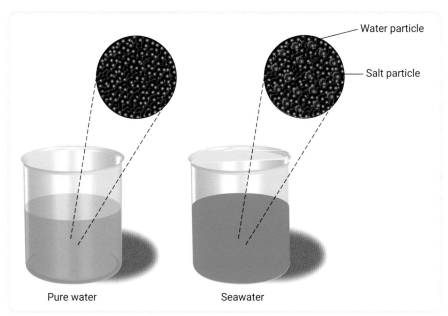

▲ FIGURE 4.1.2　In pure water all the particles are the same, making it a pure substance. Seawater is made up of different particles and so is a mixture.

It is often helpful to use simple models to represent concepts. To model pure substances and mixtures, we can represent the particles with different shapes or colours. In Figure 4.1.2, water particles are shown in blue and salt particles are shown in red. In pure water, a pure substance, all the particles are blue. Yet in the mixture, seawater, there are blue and red particles.

Classifying matter as pure substances or mixtures

Sometimes it is easy to classify a substance as a mixture because you can see that it is made up of different things. For example, the batter of chocolate chip biscuits is obviously a mixture because you can see the batter and chocolate chips. Other mixtures that are easy to classify are soda water, dirt, freshly squeezed orange juice and a meat pie.

Interactive resource
Drag and drop: Classifying mixtures and pure substances

▲ FIGURE 4.1.3　Chocolate chip biscuit batter is an example of a mixture.

▲ FIGURE 4.1.4　A cup of tea is a mixture because it is made up of water and tea.

How was it made?

Some mixtures are not easily classified just by looking at them because you cannot see different parts to them. Therefore, we need more information to help classify them as a pure substance or a mixture. One way to help classify substances is to think about how they were made. If you know different things were mixed without changing, then it will be a mixture. For example, a cup of tea looks the same throughout. Yet you know that it is a mixture because you know that it contains water as well as the tea and perhaps even sugar and milk.

Can it be separated?

Another method is to think about whether the substance can be separated. You cannot separate pure substances by simple methods, but you can separate mixtures. For example, if you heated pure water until it boiled, it is still water, but now it is a gas rather than a liquid. However, if you boiled seawater, the water would become a gas and the salt would be left behind. This means that it has been separated into the parts that made up the mixture. You will learn more about the methods that are used to separate a mixture in Chapter 5.

Table 4.1.1 lists some pure substances and mixtures.

▼ TABLE 4.1.1 Examples of pure substances and mixtures

Pure substance	Mixture
Distilled water	Tap water
Sugar	Air
Salt	Sand
Copper	Self-raising flour
Oxygen	Brass

ACTIVITY 1

Research activity: Alloys

An alloy is a mixture in which the main component is a metal. The properties of the alloy are different from the properties of the pure metal, making it useful for different purposes. For example, pure gold is very soft. However, an alloy of gold is harder, making it suitable for jewellery.

What to do

1 Conduct research about the alloys of bronze, brass, 18K gold, sterling silver, stainless steel and cast iron. For each one, find out what:
   a pure substances are in the mixture.
   b properties are different in the alloy compared with the pure metal.
   c the alloy is used for.
2 Summarise the information in a table.

▲ FIGURE 4.1.5 This statue is made from bronze, which is an alloy.

Teaching about pure substances and mixtures

★ ACTIVITY 2   4.1

A lot of research has been done to understand the best method to learn, and remember, new information. Figure 4.1.6 summarises the effectiveness of different learning methods. It shows that teaching others is the most effective learning method because students can remember nearly 90 per cent of what they learn by teaching others. However, they can only remember 5 per cent of what they learn by listening to a lecture.

You need
- distilled water
- solid copper sulfate
- 250 mL beaker
- stirring rod
- spatula

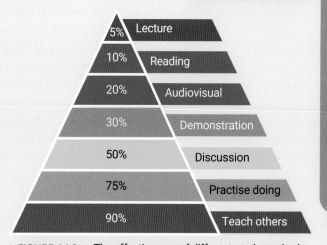

▲ FIGURE 4.1.6 The effectiveness of different study methods: teaching others is more effective than listening a lecture.

What to do

1. Pour approximately 100 mL of distilled water into the beaker.
2. Take a photo of the water and make a list of its properties.
3. Take a photo of the solid copper sulfate and make a list of its properties.
4. Use the spatula to add a small amount of copper sulfate to the water. Use the stirring rod to mix the two substances together.
5. Take a photo of the mixture and make a list of its properties.
6. Use your photos to create a slide show to teach others the difference between a pure substance (the distilled water and the copper sulfate) and a mixture (the copper sulfate in the water).
7. Show your presentation to your parents to find out how clearly and accurately you explained the ideas.

4.1 LEARNING CHECK

1. Can pure substances be separated by simple methods?
2. Give two examples of a:
   a. pure substance.
   b. mixture.
3. **Explain** why it may be difficult to classify something as a pure substance or a mixture.
4. During a science class, Amelia poured 100 mL of lemonade into a beaker. She then heated the beaker over a Bunsen burner for 15 minutes. A sugary syrup was left behind. Based on Amelia's observation, is lemonade a pure substance or a mixture? **Justify** your answer.
5. Draw a diagram to **model** a:
   a. mixture made of equal amounts of three different substances.
   b. pure substance.
   c. a mixture made of a small amount of one substance in a large amount of a second substance.
6. Your teacher will provide you with a small sample of seawater. Leave this in a small shallow dish for a few days until the water has evaporated. Take a photo of the seawater at the start and the end. Refer to the photos to **justify** why seawater is classified as a mixture.

4.2 Pure substances: going further

BY THE END OF THIS MODULE, YOU WILL BE ABLE TO:
- ✓ describe and compare elements and compounds
- ✓ give examples of elements and compounds
- ✓ classify a pure substance as an element or a compound.

> **GET THINKING**
>
> Have you heard water being described as H_2O? What do you think 'H_2O' means? Does this mean that water isn't a pure substance? Discuss these questions with a partner in a 30-second buzz. As you work through the module, think back to your discussion. Were you correct?

atom
the smallest part of an element that contains the properties of that element

element
a pure substance made up of only one type of atom; it cannot be broken down into a simpler substance

compound
a pure substance whose particles are made up of two or more different atoms chemically bonded together

Particles in pure substances

You now know that a pure substance is made up of only one type of particle. But did you know that these particles can be made up of smaller structures called **atoms**? In Year 8, you will learn a lot more about atoms. In this chapter, we are only going to consider whether the atoms are the same or not.

Elements

An **element** is a pure substance whose particles are made up of only one type of atom. Oxygen, hydrogen, nitrogen and neon are all examples of elements.

Compounds

A **compound** is a pure substance whose particles are made of different types of atoms held together by chemical bonds. Some compounds have particles that only have two atoms joined together while other compounds have particles made of hundreds of atoms.

Water is an example of a compound. Each particle in water is made up of three atoms – two atoms of hydrogen (H) and one atom of oxygen (O). This is why water is referred to as H_2O.

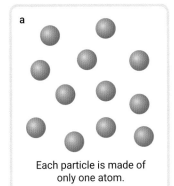

a

Each particle is made of only one atom.

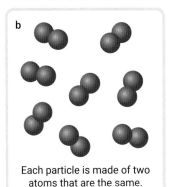

b

Each particle is made of two atoms that are the same.

▲ **FIGURE 4.2.1** The particles of elements can consist of (a) one atom, such as neon, or (b) more than one atom, such as oxygen.

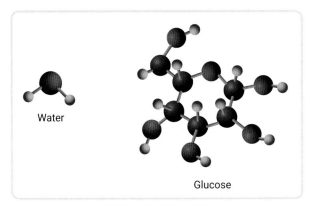

Water

Glucose

▲ **FIGURE 4.2.2** Models of particles of the compounds water (H_2O) and glucose ($C_6H_{12}O_6$). You can see that compounds have more than one type of atom in each particle.

Putting it together

In summary, all matter can be classified as a pure substance or a mixture. Pure substances can be further classified into elements or compounds.

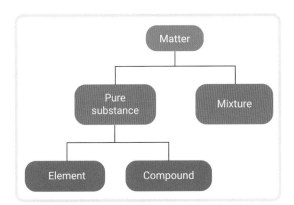

◀ **FIGURE 4.2.3** A tree diagram showing the classification of matter

4.2 LEARNING CHECK

1 **Define**:
   a element.
   b compound.

2 **Classify** each of the following as an element, a compound or a mixture.

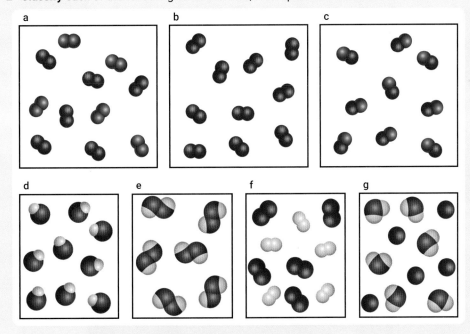

3 **State** one similarity and one difference between:
   a elements and compounds.
   b mixtures and compounds.

Video activity
Elements and compounds

Interactive resources
Simulation: Build a compound
Label: Compound, element or mixture?

4.3 Solubility

BY THE END OF THIS MODULE, YOU WILL BE ABLE TO:
- ✓ define soluble, insoluble and solubility
- ✓ give examples of soluble substances and insoluble substances
- ✓ classify substances as soluble or insoluble.

GET THINKING

Speed challenge! How many things can you list in 30 seconds that:
a can dissolve?
b cannot dissolve?

Compare your list with that of the person sitting next to you. Did they have some that you hadn't considered?

dissolve
when a substance is mixed with another and the particles from both substances spread out evenly until they are too small to see

soluble
can dissolve in another substance

insoluble
cannot dissolve in another substance

When you put a teaspoon of sugar in water, it seems to disappear – it has **dissolved** in the water. However, when you put a teaspoon of Milo in milk, only some of it dissolves.

When something dissolves, it looks as though it has disappeared. In fact, it is still there, but it is so spread out that we can't see it anymore. If something can dissolve, it is described as **soluble**. If it cannot dissolve, it is described as **insoluble**. Whether something is soluble or not depends on factors such as what the substance is, what it is being added to and the temperature.

Describing something as soluble or insoluble is a qualitative description. We can also describe it quantitatively by stating the amount that can dissolve in a certain volume. This is known as the **solubility** of a substance. For example, at 25°C, the solubility of table salt is 360 g L^{-1}, meaning that 360 grams of table salt can dissolve in 1 litre of water. In comparison, the solubility of sugar is 2000 g L^{-1}. Therefore, 2000 grams (2 kg) of sugar can dissolve in 1 litre of water.

▲ FIGURE 4.3.1 Only some of the Milo dissolves in milk.

solubility
how much of a substance can dissolve in a certain volume of another substance

◀ FIGURE 4.3.2 (a) When a solid is soluble in water (e.g. table salt) it will dissolve, and we cannot see the solid any more. (b) When a solid is insoluble (e.g. calcium carbonate), it will not dissolve in water and so we can still see the solid.

4.3 LEARNING CHECK

1 **Define**:
   a soluble.
   b insoluble.
2 **Create** a table with the column headings 'Soluble' and 'Insoluble'. In your table, **list** three things in your house that are soluble and three things that are insoluble.
3 **Classify** each of the following as soluble or insoluble.
   a Iron nail
   b Washing detergent
   c Instant coffee
   d Wool
   e Carbon dioxide (you may need to conduct some research for this one)
4 When we design products, it is important that we choose appropriate materials to make them. What do you think would happen if a car windscreen was made from a soluble substance?
5 The graph in Figure 4.3.3 shows the solubility of different substances at different temperatures. Use this information to answer the following questions.

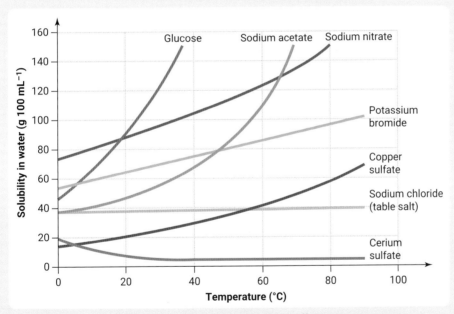

▲ FIGURE 4.3.3 The solubility of different substances at different temperatures

   a **Describe** what happens to the solubility of glucose as the temperature increases.
   b **Name** a substance that becomes less soluble as the temperature increases.
   c Your friend is making play dough and asks you to heat some water so that more salt will dissolve. Will heating the water allow more table salt to dissolve? **Explain** your answer
   d What mass of sodium acetate can dissolve in 100 mL of water if the temperature is 60°C?

Interactive resource
Drag and drop: Solubility

Extra science investigation
Design your own solubility investigation

Chapter 4 | Pure substances and mixtures

4.4 Solutions

BY THE END OF THIS MODULE, YOU WILL BE ABLE TO:
- ✓ define solute, solvent, solution, saturated, unsaturated, supersaturated, concentrated solution and dilute solution
- ✓ describe the structure and properties of solutions
- ✓ classify solutions as saturated, unsaturated or supersaturated
- ✓ classify solutions as concentrated or dilute.

Video activity
Solutions

Interactive resource
Simulation: Concentration of solutions

Extra science investigation
How much sugar?

GET THINKING

This module contains a lot of new terms! Create flashcards for the key terms. Choose how you will make your flashcards; for example, on paper or card by using an app such as Quizlet. Test yourself with your flashcards until you remember the terms and their definitions. This will make it easier to understand the concepts as you learn them in class.

Classifying mixtures

In Module 4.1, you learned how all matter can be classified as either a pure substance or a mixture. And in Module 4.2, you learned about classifying pure substances as elements or compounds. In Modules 4.4–4.6, you will learn about the different types of mixtures – solutions, suspensions and colloids.

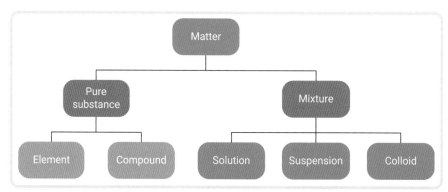

▲ **FIGURE 4.4.1** A tree diagram for the classification of matter, including types of mixtures

The structure of a solution

You already know that when a sugar cube is added to a cup of tea and stirred, the small white solid sugar grains seem to vanish as they dissolve in the water. When a substance dissolves, it joins to the other substance and spreads out evenly. You can no longer see the original substance because it has separated into particles that are so small that you can't see them. The substance that dissolves, such as sugar, is called the **solute**. The substance that does the dissolving, such as water, is called the **solvent**. Together, the solute and the solvent form a special mixture known as a **solution**.

In a sugar–water solution, the sugar is the solute because it is the substance that dissolves. The water is the solvent because it does the dissolving. The sweetened water that is produced is the solution. Similarly, a cup of coffee (the solution) is produced when instant coffee (the solute) dissolves in the water (the solvent).

solute
a substance that dissolves in another substance to form a solution

solvent
a substance that dissolves another substance to form a solution

solution
a mixture formed when a solute dissolves in a solvent

▲ FIGURE 4.4.2 (a) The solute (solid coffee) and solvent (water); (b) the solution (coffee drink)

Water is a common solvent. It is often called the universal solvent because so many substances dissolve in it. However, not everything dissolves in water. For example, nail polish does not dissolve in water, but it does dissolve in another solvent called acetone. Other liquid solvents are ethanol, kerosene and turpentine.

transparent
see-through

You are already familiar with many solutions. A salt solution is formed when salt (a solid solute) dissolves in water (a liquid solvent). Vinegar is a solution made by dissolving acetic acid (a liquid solute) in water (a liquid solvent). Soda water is made when carbon dioxide (a gas solute) is dissolved in water. We know there is a gas dissolved in soda water because, when we open the bottle, we see bubbles and hear some of the gas escape.

Appearance of solutions

Solutions are usually **transparent**, but this does not mean they cannot be coloured. If blue food colouring is added to water, you can still see through the solution. It can be described as blue (its colour) and transparent (because we can see through it).

▲ FIGURE 4.4.3 Solutions are usually transparent and may be coloured.

Concentration

Solutions can be described as **concentrated** or **dilute**. This refers to the amount of solute that is in the solution. A concentrated solution has a large amount of solute in a certain volume of the solution. By contrast, a dilute solution has only a small amount of solute in the same volume of the solution.

The terms 'concentrated' and 'dilute' are referred to as qualitative data. This is because they are a comparison or description only, not a measured amount. If we measure the amount of solute in the solution, we can calculate the **concentration**. This is often measured in the number of grams of the solute in a litre of solution. Because concentration is a measured amount, it is classified as quantitative data.

concentrated
has a large amount of solute in a certain volume of solution

dilute
has a small amount of solute in a certain volume of solution

concentration
the amount of solute dissolved in a certain volume of solution, often measured in grams per litres

In a coloured solution, the intensity of colour changes with concentration. The solute gives the solution its colour. Therefore, if there is more solute in a solution, then it will be a darker colour.

▲ **FIGURE 4.4.5** (a) A dilute cup of tea will be lighter in colour than (b) a more concentrated cup of tea.

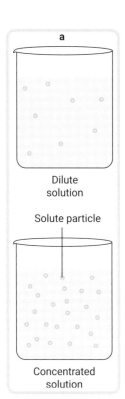

▲ **FIGURE 4.4.4** A dilute solution contains fewer solute particles than in the same volume of a concentrated solution.

saturated solution
a solution that has the maximum amount of solute dissolved in the solvent

unsaturated solution
a solution that can dissolve more solute

supersaturated solution
a solution that contains more solute than is normally able to dissolve in it at a certain temperature

Saturation

Another physical property of solutions is saturation. This is a qualitative description that refers to how much solute is dissolved in a solvent compared with the solubility (how much can dissolve).

- A **saturated solution** has the maximum amount of solute dissolved in the solvent. If more solute is added to a saturated solution, it cannot dissolve and therefore will remain as a solid.
- An **unsaturated solution** can have more solute dissolve in it. Therefore, if more solute is added to an unsaturated solution, it will dissolve.
- **Supersaturated solutions** are a special case. In most cases, if a solution is heated, more solute can dissolve than at the lower temperature. If this solution is slowly cooled, all of the solute may remain dissolved. The solution produced is described as supersaturated because it contains more solute than is normally soluble. If more solute is added to a supersaturated solution, it will disturb the solution and the extra dissolved solute can no longer stay dissolved. This means that the solute that was added, plus the extra solute that was dissolved, will appear as a solid.

ACTIVITY

Making a saturated sugar solution

Aim

To make a saturated solution of sugar and water

You need
- 100 mL water
- white sugar
- 250 mL beaker
- spatula
- stirring rod
- electronic balance

What to do

1. Pour the 100 mL of water into a beaker and place it on the electronic balance.
2. Zero the electronic balance.
3. Slowly add sugar to the water, stirring until all of the sugar has dissolved.
4. Repeat step 3 until no more sugar can dissolve. Be careful – as you get to the end, the sugar dissolves very slowly, so don't stop too early!

What do you think?

1. How much (in grams) sugar dissolved in the water?
2. What type of solution did you make (saturated, unsaturated or supersaturated)? Explain how you knew.
3. Challenge: Calculate the concentration of sugar in grams per litre.

4.4 LEARNING CHECK

1. **Match** each term with its definition.

Term	Definition
Concentrated solution	A solution with a small amount of solute in the solution
Unsaturated solution	A solution with the maximum amount of solute dissolved in the solution
Dilute solution	The substance that does the dissolving
Solute	A solution containing more than the maximum amount of solute in the solution
Saturated solution	A solution with a large amount of solute in the solution
Supersaturated solution	What is produced when a solute dissolves in a solvent
Solvent	A solution that can still dissolve more solute
Solution	The substance that is dissolved

2. **Describe** what happens to the particles of a solute when they dissolve in a solvent.
3. Reflect on the way a solution is made and what solutions look like. **List** four solutions in your home. For each one, **state** the solute and solvent and **describe** its appearance.
4. **Explain** why a saturated solution has a greater concentration than an unsaturated solution of the same solute.
5. When you get wet from the rain, you may describe yourself as saturated. **Compare** the use of the term 'saturated' in this example to its use in describing solutions.
6. **Discuss** whether 'weak' and 'strong' are the correct scientific terms to describe drinks of cordial that have different amounts of flavouring.
7. Challenge: A solution contains 5 g of sodium carbonate dissolved in 2 L of water. **Calculate** the concentration of the solution in grams per litre.

4.5 Suspensions

BY THE END OF THIS MODULE, YOU WILL BE ABLE TO:
- ✓ define suspension
- ✓ describe the structure and properties of suspensions.

Quiz
What is a suspension?

GET THINKING

Think about what happens when dirt is added to a bucket of water.
a Does the dirt dissolve?
b What happens if you leave it for an hour?
c What do you think happens to the dirt particles?
d Why do you think this happens?

The structure and properties of a suspension

Not all mixtures are solutions. If the substance being added is not soluble, then it will not dissolve. This means that the mixture that is produced is not a solution. Instead, a **suspension** or a colloid is formed. In this module, you will learn about suspensions.

A suspension is formed when an insoluble substance is added to the solvent. Although it may initially spread throughout the solvent, if it is left for a while the substance settles to the bottom of the container.

Adding dirt to water produces a suspension. Initially, the water looks dirty throughout because the dirt particles are mixed with the water. However, if the mixture is left, the dirt will settle to the bottom. This means that the suspension forms layers.

suspension
a mixture of at least one insoluble solid and a liquid or solution, where the insoluble substance settles to the bottom of the container over time

▲ **FIGURE 4.5.1** (a) Powdered chalk added to water and mixed will form a suspension. (b) If the suspension is left undisturbed, the chalk will settle to the bottom of the beaker.

Sometimes it is difficult to know whether the mixture is a solution or a suspension. The key difference is that a solution is usually transparent because the substance has dissolved. In a suspension, the substance did not dissolve and therefore it can still be seen. This means that when the suspension is mixed, it is not transparent. A second way to tell the difference is that a suspension will settle into layers, but a solution will not. This is because the substance in a suspension is too heavy to stay **suspended** in the liquid.

suspend
hang or keep from falling

Understanding the appearance of beach water in different weather conditions

★ ACTIVITY 4.5

On a calm day, the water at the beach can be crystal clear, as in Figure 4.5.2. However, when the weather is windy, the water can be opaque and dirty as in Figure 4.5.3.

▲ FIGURE 4.5.2 Crystal clear ocean water

▲ FIGURE 4.5.3 Dirty beach water

We can use our understanding of the structure and properties of suspensions to explain the beach water in each of these situations.

1 Brainstorm the reasons for the different appearances of the beach water. Use the template in Figure 4.5.4 to create a mind map with your ideas.

2 Make a model of the beach water by putting some sand and water in a large beaker. How will you replicate calm weather or rough weather?

3 Use your brainstorm, model and information in this module to write a paragraph explaining the appearance of beach water in different weather conditions.

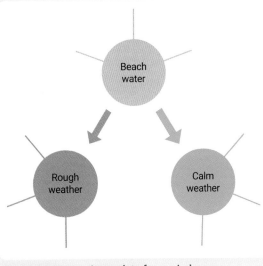

▲ FIGURE 4.5.4 A template for a mind map

4.5 LEARNING CHECK

1 **Define** 'suspension'.
2 **State**, and **justify**, the property that determines whether a substance will form a suspension, and not a solution, when added to water.
3 **List** two examples of suspensions.
4 Calcium carbonate is insoluble in water. During an experiment to observe mixtures, Tyrique added a spatula of calcium carbonate powder to water in a beaker. The mixture was stirred with a stirring rod and then left undisturbed for an hour. **Describe** the predicted observations:
   a when the mixture was first stirred.
   b after the mixture had been left for an hour.

4.6 Colloids

BY THE END OF THIS MODULE, YOU WILL BE ABLE TO:
- ✓ define colloid, Tyndall effect, emulsion, foam, gel and aerosol
- ✓ explain the structure and properties of colloids.

GET THINKING

Observe the glass of milk in Figure 4.6.1. Is milk a solution or a suspension or neither? List the characteristics of milk that correspond to solutions and the characteristics that correspond to suspensions. Use this to justify your answer.

The structure and properties of a colloid

A **colloid** is another type of mixture. A colloid is similar to a suspension because it forms when an insoluble substance is added to another substance. However, in the case of a colloid, the particles are so small that they do not settle to the bottom. Instead, they stay spread out in the mixture.

Ink is a colloid. It has solid pigment particles spread evenly in a liquid such as oil or water. Ink is not a solution because the solid pigment particles do not dissolve in the oil or water. It is not a suspension either because the mixture does not settle into layers. Other examples of colloids are milk, mayonnaise, smoke, paint, whipped cream and blood.

The particles in a colloid are larger than the particles in a solution and so the mixture is cloudy, or **opaque**, rather than transparent. The number of insoluble particles affects how cloudy the colloid is. A colloid with many insoluble particles is more opaque than one with a smaller number of insoluble particles.

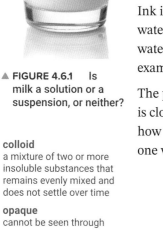

▲ **FIGURE 4.6.1** Is milk a solution or a suspension, or neither?

colloid
a mixture of two or more insoluble substances that remains evenly mixed and does not settle over time

opaque
cannot be seen through

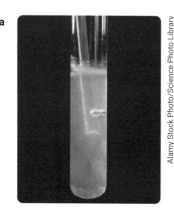 

▲ **FIGURE 4.6.2** Test tubes of colloids. **(a)** This test tube has fewer insoluble particles than **(b)** this test tube and so is less opaque.

Tyndall effect
when the insoluble particles in a colloid interact with a beam of light to enable you to see the beam of light

It is possible to identify whether a mixture is a solution or a colloid by shining a thin beam of light through it. In most solutions, you will not be able to see the beam of the light. In a colloid, the insoluble particles will interact with the light, which means you will be able to see the beam of light. This is called the **Tyndall effect** (Figure 4.6.3).

Types of colloids

We can further classify colloids by looking at the different states of the substances that are mixed.

- **Foams** are formed when a gas is spread through a liquid; for example, whipped cream or shaving cream.
- **Emulsions** are a type of colloid containing only liquids. Both cream and homogenised milk are stable emulsions. They are cloudy white mixtures with liquid milk fat mixed evenly throughout liquid water. Cream is more opaque than milk because there is more milk fat mixed throughout the liquid water.
- **Gels** are formed when a solid is spread throughout a liquid; for example, jelly, paint or blood.
- **Aerosols** are formed when a solid or a liquid is spread through a gas. Hairspray is an aerosol that has a liquid spread through a gas. Smoke has solid ash particles spread through a gas (the air).

▲ **FIGURE 4.6.3** A beam of light is not visible through the solution on the left, but is visible through the colloid on the right. This is known as the Tyndall effect.

foam
a type of colloid in which a gas is mixed in a liquid

emulsion
a type of colloid in which a liquid is mixed in another liquid

gel
a type of colloid in which a solid is mixed in a liquid

aerosol
a type of colloid in which a solid or a liquid is mixed in a gas

4.6 LEARNING CHECK

1. **Copy and complete** the table by **recalling** the components of different types of colloids.

2. **Describe** one method of differentiating (telling the difference between):
   a. a colloid and a solution.
   b. a colloid and a suspension.

3. **Classify** each of the following as solution, suspension or colloid.
   a. Soft drink
   b. Sand mixed in water
   c. Pancake batter
   d. The air near a bushfire
   e. Ocean water

4. **Discuss** whether the orange juice in Figure 4.6.4 is a solution, a suspension or a colloid. Justify your answer by referring to the properties of each of the types of mixtures.

5. Milk is a colloid. You can see the Tyndall effect for yourself if you add some drops of milk to water, darken the room and then shine a torch through the glass from the side. Take a photo that shows the Tyndall effect. Annotate your photo to **explain** why it occurs.

Type of colloid	What is spread in the colloid	What it is spread in
Foam		
	Solid	Gas
Aerosol	Liquid	
	Liquid	Liquid
Gel		

Interactive resource
Crossword: Solutions, suspensions and colloids

▲ **FIGURE 4.6.4** Is freshly squeezed orange juice a solution, a suspension or a colloid?

4.7 Air quality

SCIENCE AS A HUMAN ENDEAVOUR

BY THE END OF THIS MODULE, YOU WILL BE ABLE TO:
- explain how scientists use the concentration of pollutants as a measure of air quality.

Video activity
Measuring air quality

▲ **FIGURE 4.7.1** Air pollution is a serious health issue.

Air is a mixture of gases such as nitrogen, oxygen, carbon dioxide, neon and hydrogen. Figure 4.7.2 shows the percentage of each gas normally found in air.

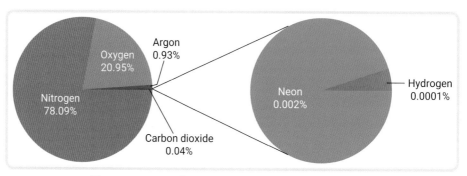

▲ **FIGURE 4.7.2** The percentages of gases in air

Unfortunately, there are also substances in the air that can be harmful. These are called **pollutants**. Some common pollutants are:
- carbon monoxide from vehicles and industry
- sulfur dioxide from power plants and industry
- nitrogen dioxide from vehicles, industry and gas heaters
- small particles in smoke from wood fire heaters and bushfires
- dust
- pollen.

pollutant
a substance introduced into an environment that can be harmful

It is estimated that, during 2020, air pollution caused more than 163 000 deaths in the world's largest cities, including Tokyo (Japan), Delhi (India), Shanghai (China), Mexico City (Mexico) and São Paulo (Brazil). Scientists have developed the **air quality index** (AQI), which indicates the level of pollutants in the air. An AQI of less than 50 indicates that the level of pollution is low and is unlikely to affect most people's health. The higher

air quality index
a measure of the level of pollution in the air, expressed as a number from 0 (no pollution) to 500 (maximum pollution)

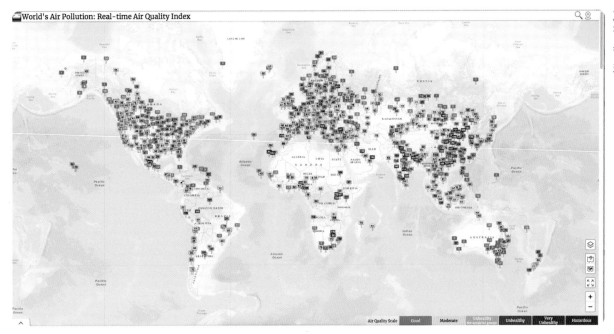

▲ FIGURE 4.7.3 The AQI across the world in April 2022. Most places in Australia register as good or moderate on the scale.

the AQI, the higher the pollution levels and the the greater the risk of harm to people's health.

By measuring the concentration of pollutants, scientists can monitor and communicate risks to the community. Many weather apps include the local AQI. Some government websites also report the AQI and provide advice for residents. Some websites also have a pollen forecast', which indicates how much pollen there is likely to be in the air. This means that people at higher risk (e.g. those with respiratory diseases) can take precautions such as staying inside. During windy spring days or when there are bushfires nearby, people should close windows and doors to keep pollutants such as pollen and smoke out of the house. Similarly, in Australia it is now illegal to smoke cigarettes in a car when there is a child present because the cigarette smoke, a pollutant, is a risk to the child.

Weblink
World's Air Pollution: Real time Air Quality Index

▲ FIGURE 4.7.4 The AQI is included in many weather apps.

4.7 LEARNING CHECK

1 What does AQI stand for?
2 List the following gases in order from the highest percentage in air to the lowest: carbon dioxide, oxygen, nitrogen, argon, neon, hydrogen.
3 **List** five pollutants that are commonly found in air.
4 **Explain** why air is classified as a mixture.
5 Amy suffers from asthma. She regularly checks the AQI for her suburb, especially when it is smoky outside. **Explain** how the AQI could help Amy avoid an asthma attack.
6 The AQI for Yinchuan in China was 291 on 8 January 2022. **Explain** why it would be advisable to wear a mask when outside.

SCIENCE INVESTIGATIONS

4.8 Using observations to make classifications

SCIENCE SKILLS IN FOCUS

IN THIS MODULE, YOU WILL FOCUS ON LEARNING AND IMPROVING THESE SKILLS:

- using observational skills to make classifications.

Observations play a key role in science because they are data that can be analysed and interpreted. Observations can be:

- qualitative if they are descriptive only. We use our senses to collect qualitative data – what we see, hear, feel, taste or smell
- quantitative if they are measured. You have already learned about measuring mass, temperature, volume, length and time. Other measurements include concentration, pH and pressure.

Video
Science skills in a minute: Observations

Science skills resource
Science skills in practice: Observing to make classifications

OBSERVING MIXTURES

AIM

To observe different substances and use the observations to make appropriate classifications

YOU NEED

- ☑ test tubes
- ☑ test-tube racks
- ☑ plastic droppers
- ☑ approximately 50 mL water
- ☑ small spatula or spoon
- ☑ stoppers for test tubes
- ☑ variety of substances, such as table salt, jelly crystals, methylated spirits, vinegar, olive oil, kerosene, sand, powdered chalk, milk, cream, small pieces of cork, tea leaves

WHAT TO DO

In a group of two or three:

1. Construct a results table similar to Table 4.8.1. Your teacher will tell you what substances you will be testing – list these in the column with the heading 'Substances'.

2. Place the test tubes in the test-tube rack and label each with the name of one of the substances.

3. Add a small amount of each substance to the relevant test tube. For solids, add about a spatula tip full. For liquids, use a clean plastic dropper to add liquid to a height of about 2 cm in the test tube.

4. Observe each substance, recording the physical properties in the results table.

5. Use a plastic dropper to add enough water to half-fill each test tube.

6. Place a stopper on each test tube and carefully shake it from side to side for 1 minute.

7. Observe each mixture, recording your observations in the results table.

8. Place each mixture in the test-tube rack and leave for 15 minutes.

9. Observe each mixture, recording your observations in the results table.

Warning

Broken glass can cut skin. Use test-tube racks to hold test tubes. Clean up any broken glass immediately and put it in the glass bin.

Substances can get into your eyes and cause damage. Always wear safety glasses.

Some of the substances used in this experiment are flammable. Perform the experiment away from naked flames.

Some of the substances, such as kerosene, can be irritating to skin and respiratory surfaces. Wear protective clothing. Perform the experiment in a well-ventilated area.

RESULTS

Record your observations in your results table (Table 4.8.1).

▼ TABLE 4.8.1 Observations of different mixtures

Substance	Observations before water was added	Observations as soon as water was added	Observations after 15 min	Type of mixture

WHAT DO YOU THINK?

1. **Classify** each of the mixtures as a solution, suspension or colloid. **Record** your classification in the results table.

2. **Compare** your classifications with other groups in the class. Where there any differences? If there were, suggest a reason why.

3. **Explain** how your observations allowed you to classify each mixture.

4. **List** the variables in this experiment. Which of the variables were controlled? Which were not controlled?

5. **Explain** why the solutions were left for 15 minutes.

6. **Discuss** the concepts of pure substances and mixtures with reference to the mixtures you have made.

CONCLUSION

Write one or two sentences summarising the importance of making careful observations when classifying substances.

4 REVIEW

REMEMBERING

1 **Write** the term that matches each description.
   a A pure substance whose particles have more than one type of atom
   b A mixture that is usually transparent
   c The substance that is dissolved in a solution
   d A pure substance whose particles are made up of only one type of atom
   e A mixture that separates into layers
   f The substance that does the dissolving in a solution
   g A mixture that allows a beam of light to be visible

2 **State** the term given to the physical property of being able to see through something.

3 **State** the phases of the substances that make an emulsion.

4 **Describe** the Tyndall effect and identify its use in classifying mixtures.

5 **State** whether each of the following statements is true or false. Rewrite each false statement to make it true.
   a Solutions are usually transparent.
   b Solutions are always colourless.
   c Fresh orange juice containing pulp is a solution.
   d A solution always has water in it.
   e When acetone removes nail polish, the nail polish is the solute and acetone is the solvent.
   f If soda water goes flat, it is no longer a solution. (You may need to research what soda water is made up of to answer this question.)
   g The particles in a colloid are smaller than in a suspension but larger than in a solution.
   h Water-soluble paint is a solution.

6 **Describe** the structure, at a particle level, of a:
   a pure substance.
   b mixture.

UNDERSTANDING

7 **Explain** why a suspension will separate into layers if left long enough, but a colloid won't.

8 **Explain** two ways you could tell the difference between a solution, a suspension and a colloid in the laboratory.

9 A teaspoon of salt was added to a glass of hot water and stirred.
   a **Describe** what you would observe.
   b **Explain** why this occurs.
   c **State** the:
      i type of mixture the salt and water forms.
      ii name of the salt.
      iii name given to the water.
   d If you kept adding more teaspoons of salt to the glass of water, do you think the salt would continue to dissolve? **Explain** your answer.

10 **Identify** whether each of the following statements is true or false. If it is false, **explain** why it is false.
   a In a suspension, particles always fall to the bottom of the container.
   b All emulsions are colloids.
   c All colloids are emulsions.
   d A suspension cannot contain a solution.

APPLYING

11 During an experiment, Adam forgot to label his beakers of a blue solution produced when he dissolved copper sulfate in water. Adam knows that one solution is unsaturated, and the other is saturated. **Describe** two different methods that he could use to correctly identify each solution and the observations that he would make for each beaker. **Explain** why each observation occurred.

12 Use the words 'dilute' and 'concentrated' to **describe** the two glasses of cordial shown.

13 Is a lake an example of a suspension? **Explain** your answer.

14 To make a simple form of tempera paint, egg yolk is added to a mixture of solid pigment and oil.
   a **Predict** which type of colloid the paint is.
   b **Suggest** a reason why the egg yolk is needed.

EVALUATING

15 Ming is testing three pure substances (A, B and C), which are all solids. She prepares three beakers, each with 100 mL of water at the same temperature. Ming adds 50 g of substance A (in 10 g amounts) to the first beaker and stirs the mixture. She repeats this for the other solids and beakers. Ming's observations are shown in the table. **Classify** the types of mixtures she created when she added A, B and C to the water. **Explain** the reasons for each classification.

Substance	Observations when mixed with 100 mL of water
A	As each 10 g amount is added, it dissolves instantly. After all the solid has been added, it is no longer visible. The mixture is transparent.
B	As each 10 g amount is added, the solid floats in the water. After 20 minutes, the solid has settled to the bottom of the container. None of the solid has dissolved.
C	Each 10 g amount seems to mix immediately. The mixture is not transparent. It does not separate into layers. After 20 minutes, there is still only one cloudy layer.

16 The solubility of potassium nitrate (KNO_3) is shown in the following graph.
   a How many grams of potassium nitrate can dissolve in 100 g of water at 60°C?
   b If 50 g of potassium nitrate was stirred into 100 g of water at 40°C, would the solution be unsaturated, saturated or supersaturated? **Justify** your answer.
   c What mass of potassium nitrate should be added to 200 g of water at 20°C to produce a saturated solution? **Explain** the reason for your answer.
   d If 150 g of potassium nitrate needs to be dissolved in 100 g of water, what temperature should the water be heated to?

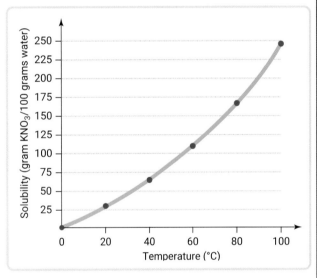

17 The saying 'clear as mud' is used when you try to explain something to someone, but they do not understand straight away.
   a **State** what type of mixture mud is.
   b How does this saying suggest that you do not understand what is being explained?
   c Using a different mixture, **write** a saying you could use if you understand an explanation straightaway. **Explain** how your saying achieves this.

CREATING

18 Construct a question that has the answer of:
   a dissolve.
   b soluble.
   c solute.
   d solution.
   e solubility.
   f dilute.
   g unsaturated.

19 In this chapter, you have learned about a lot of different ways of classifying mixtures. **Create** a concept map (a mind map) to organise the information. Include the key terms in boxes or circles, using colours to show similarities or differences. Join the boxes or circles with linking lines to show the connection between terms. Complete your concept map by adding notes for definitions, diagrams and examples. The following diagram shows a simple structure for a concept map.

20 When you make a glass of cold Milo, you usually end up with a glass of chocolate-flavoured milk and a layer of solid Milo on top. **Create** an explanation of the composition of a glass of cold Milo by using the terms 'matter', 'pure substance', 'mixture', 'solution', 'suspension' and 'colloid'.

▲ A glass of cold Milo

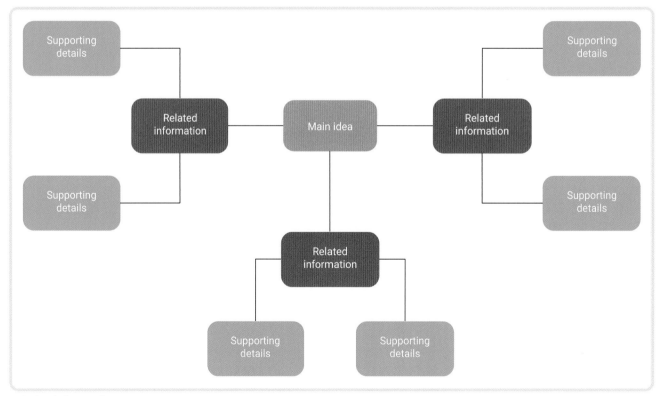

▲ A simple format for a concept map

BIG SCIENCE CHALLENGE PROJECT #4

Just like us, fish need oxygen. However, unlike us, fish can get their oxygen from water. Some oxygen from the air dissolves in the water, and fish have gills that can extract the oxygen from the water.

The amount of oxygen dissolved in the water varies depending on conditions such as temperature, aeration and salinity. People who keep fish in an aquarium or a pond need to monitor the levels of dissolved oxygen to make sure that there is enough oxygen for the fish.

1. Connect what you've learned

Write a paragraph explaining what dissolved oxygen in an aquarium is. Use relevant scientific terminology such as 'pure substance', 'mixture', 'solution', 'solute', 'solvent' and 'concentration' in your explanation.

2. Check your thinking

Conduct research to learn about the effect temperature, aeration and salinity have on levels of dissolved oxygen. Summarise your findings in a table similar to the one below.

▼ The effect of environmental factors on dissolved oxygen levels

Factor	Does dissolved oxygen increase or decrease when the factor is increased?	Does dissolved oxygen increase or decrease when the factor is decreased?
Temperature		
Aeration		
Salinity		

3. Make an action plan

What steps do people with aquariums or ponds take to increase the levels of dissolved oxygen?

4. Communicate

Create a poster to outline how to maintain an appropriate level of dissolved oxygen in an aquarium or pond. The poster should be suitable for people who have never owned fish before, so it needs to be clear and interesting.

▲ Fish have gills to extract oxygen from water.

5 Separating mixtures

5.1 Physical properties (p. 124)
Substances in a mixture have different physical properties.

5.2 Separating suspensions: sedimentation, centrifuging and decanting (p. 127)
Sedimentation, centrifuging and decanting are different ways of separating suspensions.

5.3 Separating suspensions: filtration (p. 131)
Filtration separates a suspension on the basis of particle size.

5.4 Separating solutions: evaporation, crystallisation and distillation (p. 134)
Evaporation, crystallisation and distillation are different ways of separating solutions.

5.5 Other separation techniques: magnetic separation, flocculation and chromatography (p. 139)
Flocculation causes small particles to clump together, allowing them to settle and be removed. Any difference in properties, such as magnetism, mass, charge and solubility, can be used to separate a mixture.

5.6 FIRST NATIONS SCIENCE CONTEXTS: Separating techniques traditionally used by First Nations Australians (p. 142)
For thousands of years, First Nations Australians have used a variety of techniques to separate out parts of mixtures.

5.7 SCIENCE AS A HUMAN ENDEAVOUR: Separation techniques used in recycling (p. 145)
During recycling, materials are broken down and made into new products.

5.8 SCIENCE INVESTIGATIONS: Writing a method (p. 146)
Writing a method to separate a mixture

BIG SCIENCE CHALLENGE #5

▲ FIGURE 5.0.1 Accidentally mixing staples, sugar and flour

Imagine that you were cleaning out your pencil case in the kitchen and accidentally knocked your staples into the sugar bowl, which fell, tipping into some flour that was ready for making scones. You know that this was the last of the sugar, and so it is important to recover it before someone needs it for a cup of tea.

▶ How will you separate the sugar, flour and staples?

#5 SCIENCE CHALLENGE ACCEPTED!

At the end of this chapter, you can complete Big Science Challenge Project #5. You can use the information you learn in this chapter to complete the project.

Assessments
- Prior knowledge quiz
- Chapter review questions
- End-of-chapter test
- Portfolio assessment task: Science investigation

Videos
- Science skills in a minute: Methods **(5.8)**
- Video activities: Density of solids **(5.1)**; Centrifuging blood plasma **(5.2)**; Salt: separating mixtures **(5.4)**; Fractional distillation **(5.4)**; How recycling works **(5.7)**

Science skills resources
- Science skills in practice: Writing a method **(5.8)**
- Extra science investigations: Rate of evaporation of water **(5.4)**; Does ink contain water? **(5.5)**; Flocculation **(5.5)**

Interactive resources
- Simulation: Density **(5.1)**
- Drag and drop: What is filtration? **(5.3)**
- Label: Lab equipment used to separate mixtures **(5.4)**
- Crossword: Separating substances **(5.5)**

Nelson MindTap

To access these resources and many more, visit:
cengage.com.au/nelsonmindtap

Chapter 5 | Separating mixtures

5.1 Physical properties

BY THE END OF THIS MODULE, YOU WILL BE ABLE TO:
- ✓ define physical property, solubility, melting point, boiling point and density
- ✓ describe the physical properties of substances.

Video activity
Density of solids

Interactive resource
Simulation: Density

GET THINKING

Observe the photo of sugar and flour in Figure 5.1.1. Think about what you know about sugar and flour. Write a list of the properties of each substance – what properties do they have in common and what properties are different?

▲ FIGURE 5.1.1 How are sugar and flour similar and how are they different?

Physical properties of chemical substances

Just as people have different features, so do chemical substances. These features are called properties. There are two types of properties: **chemical properties** and **physical properties**. Chemical properties relate to how a chemical changes to produce new substances. Physical properties, such as colour and hardness, are features that do not involve a change in the chemical. This module will focus on physical properties.

chemical property
a property of a substance that shows how it reacts when combined with other substances

physical property
a property of a substance that can be observed or examined without changing the composition of the substance

In Chapter 4, you learned that a mixture contains more than one type of substance. Because these substances are not combined chemically, we can use simple methods to separate them. Each method is based on a particular physical property that is different for the substances in the mixture. This is similar to separating school students into classes on the basis of their different ages. In this module, you will learn about the different physical properties so that you can apply them in the following modules about separating mixtures.

Colour

The colour of a substance is a physical property. You may be able to use the colour of a substance to identify what it is made up of. For example, if a cake is brown, then you could infer that there is cocoa in it and therefore it is a chocolate cake. Or if your drink is yellow, then it might be lemon flavoured. Some chemicals are also coloured. Gold metal has a characteristic yellow colour, whereas copper metal is described as salmon pink. A solution of copper sulfate is blue, whereas a solution of potassium permanganate is purple–pink (Figure 5.1.2).

▲ FIGURE 5.1.2 Some solutions have distinct colours, such as copper sulfate (blue), nickel sulfate (green) and potassium permanganate (purple–pink).

124 Nelson Science 7 | Australian Curriculum

Sometimes, the colour determines what chemical is chosen for a particular purpose. First Nations Australian artists used their knowledge of the colour of substances well before modern paints were available. If the artist wanted an orange paint, they mixed ochre clay with water because it contains iron oxide, which gives it an orange colour. Or if they wanted black, the artist would use charcoal.

Transparency

Transparency refers to whether you can see through a substance. If you can see through something, it is transparent. However, if you cannot see through it, it is described as opaque.

A glass window is fully transparent because you can see through it. A brick wall is opaque because you cannot see through it. Some chemical substances, such as white vinegar, are transparent. By contrast, others, such as copper metal, are opaque.

In Chapter 4, you learned that solutions are generally transparent, but colloids are not. Therefore, the physical property of transparency may be used to identify whether a liquid is a solution or a colloid.

Soluble or insoluble

Whether something is soluble or not (insoluble) is another physical property. It could be easy to think that this is a chemical property because the solid has disappeared when it dissolves. However, it is still the same chemical, just spread out into particles that are so small we can't see them anymore. This means that being able to dissolve (or not dissolve) is a physical property.

density
the mass of a substance in a certain volume

immiscible
unable to mix; separates into layers if combined

Density

Density is how much mass is in a certain volume. In fishing, a lead sinker is used to take the fishing line down through the water. It does this because lead is dense – it is heavier than most other substances that are the same size. A marshmallow is not very dense – it is lighter than most other substances of the same size.

When two **immiscible** substances are combined, they do not mix. The less dense substance rises and sits on top while the denser substance sinks and sits on the bottom. This is why oil forms a layer on top of water – it is less dense than water (see Figure 5.1.3).

▲ FIGURE 5.1.3 Oil forms a layer on top of water because it is less dense than water. Oil and water are immiscible substances.

Melting and boiling points

In Chapter 3, you learned about changes of state. The melting point is the temperature at which a substance changes between a solid and a liquid when it is melting or freezing. Similarly, the boiling point is the temperature at which all of a substance changes between a liquid and a gas when it boils or condenses.

▲ FIGURE 5.1.4 Water at boiling point

Melting and boiling points are important physical properties because they determine the state of matter that a substance normally exist as. For example, water has a melting point of 0°C and a boiling point of 100°C. Therefore, water is a liquid at room temperature (approximately 25°C) because the temperature is high enough for the water to melt, but not high enough for it to boil.

Other physical properties

Other physical properties include:

lustre
how shiny a metal is

- **lustre** – how much a substance reflects light. If something is shiny, such as diamond, then it is described as lustrous. If a substance such as wood is not shiny, then it is described as dull

viscosity
a liquid's resistance to flowing

- **viscosity** – a liquid's resistance to flowing. Liquids that are thick and sticky, such as honey, have a high viscosity. Liquids that flow easily, such as water, have a low viscosity
- **texture** – what something feels like; for example, smooth or rough
- **magnetism** – being attracted to a magnet
- **smell** – for example, no odour, sweet smelling or pungent (a strong, sharp smell)
- **hardness** – for example, soft, brittle or hard.

5.1 LEARNING CHECK

1. **List** five physical properties of water.
2. **Explain** why taste is not included in the physical properties of chemicals in a science laboratory.
3. **Compare** the physical properties of milk and lemonade. State two physical properties that they have in common and two physical properties that are different.
4. Physical properties can be classified as extrinsic or intrinsic. An extrinsic property changes depending on the amount of the substance, whereas an intrinsic property will always stay the same. Do you think intrinsic properties or extrinsic properties are more valuable when describing chemicals? **Justify** your answer.
5. **Create** a table to summarise the description and examples for the physical properties covered in this module.

5.2 Separating suspensions: sedimentation, centrifuging and decanting

BY THE END OF THIS MODULE, YOU WILL BE ABLE TO:
- ✓ define sedimentation, centrifuging, decanting and decantation
- ✓ state the property that allows sedimentation, centrifuging and decanting to separate mixtures
- ✓ explain how sedimentation, centrifuging and decanting separate mixtures.

GET THINKING

Look at the pictures in this module. What word describes each action that is occurring? As you learn about each separation technique, relate the information to this word. By linking new information to something you already know, you are more likely to remember it!

Video activity
Centrifuging blood plasma

Sedimentation

When a suspension has been left for a while, the insoluble substance separates from the liquid. If the solid is dense enough, it settles on the bottom (Figure 5.2.1). This solid is called the **sediment** and the process is called **sedimentation**. You can see sedimentation after soaking dirty clothes. If you leave the water in the sink, the dirt will settle on the bottom, leaving clear water above it.

Centrifuging

Sedimentation can be a slow process because it relies on gravity to separate the mixture into layers. **Centrifuging** is used to quickly separate a suspension or colloid into layers based on the density of the substances.

A **centrifuge** is a machine that spins very fast, like a washing machine on its spin cycle (Figure 5.2.2.). As it spins, the denser substance moves to the sides of the container. Some centrifuges also include a filter that allows only the smaller substances through. A washing machine works in this way. As it spins, the wet clothes are pushed to the sides of the drum. The holes on the outer wall allow only the water to pass through, and so the water is separated from the clothes. In the kitchen, a lettuce spinner works in the same way. As it spins, the wet lettuce is pushed to the sides of the spinner and the water passes through the holes, leaving the lettuce dry.

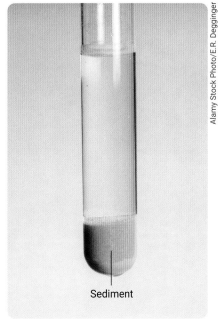

▲ **FIGURE 5.2.1** During sedimentation, a heavy sediment settles on the bottom of the container.

sediment
the insoluble solid that settles on the bottom of a suspension

sedimentation
the process of particles settling on the bottom of the liquid part of a suspension

centrifuging
the process of using a centrifuge to separate a mixture

centrifuge
a machine that spins very fast and separates heavier substances from lighter substances

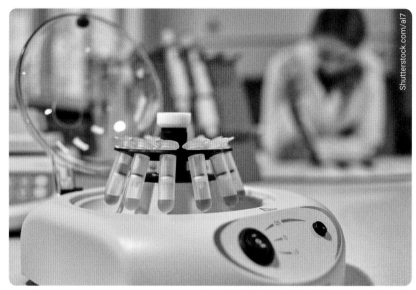

▲ FIGURE 5.2.2 A centrifuge spins very quickly to separate the mixtures in the tubes.

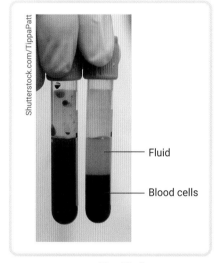

▲ FIGURE 5.2.3 Blood before centrifuging (left) and after centrifuging (right). Centrifuging separates the blood into layers.

Laboratory centrifuges hold special tubes containing the mixture. As the tubes spin, the denser substance goes to the bottom of each tube. When the spinning is finished, the mixture has been separated into layers. Doctors, veterinarians and medical scientists use this technique to separate the parts of blood samples (Figure 5.2.3). This allows them to determine how much of the blood is fluid and how much is cells.

Decantation

Decantation is the process of pouring off a liquid. Decantation can separate mixtures that form layers, such as a suspension of dirt in water, or immiscible liquids such as oil and water. Pouring off the top layer separates it from the rest of the mixture.

Decanting is often used in everyday life. For example, once you have cooked pasta, you can carefully pour the water out of the saucepan, leaving the pasta.

Decanting is not a very precise way to separate a suspension: it can stir up some of the solid particles and these may be poured out with the liquid. For this reason, the last bit of liquid is usually left in the container with the sediment and not all the solid is recovered.

In the laboratory, a stirring rod can direct the fluid as it is being poured off. This is shown in Figure 5.2.4.

decantation
the process of decanting

decanting
pouring off the top, less dense liquid

▲ FIGURE 5.2.4   Decanting a suspension

Separating a suspension by decantation

☆ ACTIVITY

You need
- dirt
- water
- two 250 mL beakers
- 100 mL measuring cylinder
- stirring rod
- teaspoon

What to do
1 Measure 100 mL of water in the measuring cylinder.
2 Pour the water into one of the beakers.
3 Add a teaspoon of dirt to the water and stir it with the spoon until it is evenly mixed.
4 Describe the mixture.
5 Leave the mixture overnight.
6 Describe the mixture the next morning.
7 Decant the water into the second beaker, using the stirring rod to direct the flow. Stop decanting before the dirt layer.
8 Measure the volume of clean water with the measuring cylinder.

What do you think?
1 Compare your descriptions of the mixture from steps 4 and 6. Explain why they are different.
2 Your mixture started with 100 mL of water. Why weren't you able to decant all of the water from the mixture?

5.2 LEARNING CHECK

1 **Define**:
   a sediment.
   b centrifuging.
   c decantation.
2 **Describe** the process of decantation.
3 **State** a disadvantage of decantation.
4 **Explain** the advantage of centrifuging over sedimentation.
5 In the carnival ride shown in Figure 5.2.5, people are strapped in around the edge of the ride, which spins very fast.

▲ FIGURE 5.2.5

   a **Describe** what the people would feel during the ride.
   b **Compare** this to a centrifuge by stating one thing that is the same and one thing that is different.
6 Having visual aids can help us remember information. In this module, you learned about some different ways of separating mixtures. **Create** a poster for each method to summarise the key information. Use a clear diagram or photo as the main part of your poster, with labels and text boxes for other facts.

5.3 Separating suspensions: filtration

BY THE END OF THIS MODULE, YOU WILL BE ABLE TO:
- ✓ define filtering, filtration, filter funnel, filter paper, filtrate and residue
- ✓ label the components of a filtration apparatus
- ✓ explain how filtration separates the components in a suspension.

GET THINKING

When you are shopping online, you may use filters to refine your search. In small groups, discuss the purpose of these filters and how they work. Summarise your discussion in a few sentences, including an example. As you work through this module, reflect on your discussion and how filtering in science is similar to filtering while online shopping.

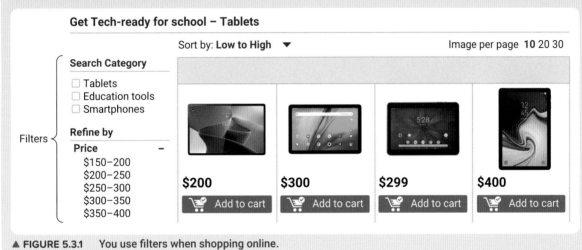

▲ FIGURE 5.3.1 You use filters when shopping online.

Filtration

Filtration is a technique that separates mixtures based on the size of the particles. When scientists are **filtering**, they use a filter such as **filter paper** with very small holes (or pores). Particles that are smaller than the holes will pass through the filter paper and form the **filtrate**. Particles that are bigger than the holes will not pass through. They are left on the filter paper and form the **residue**.

Suspensions can be separated by filtration. The solid particles are too big to pass through filter paper but the particles dissolved in the liquid are small enough to pass through. Filtration will not separate a solution or colloid because all the particles are small enough to pass through the filter paper.

We use filters every day. Vacuum cleaners contain dust bags with small pores that trap the solid dust particles but let the air pass through. Cars have air filters that remove dust and dirt so that they do not damage the engine. Clothes dryers have a filter that traps lint so that it doesn't end up all over your clothes.

filtration
a process used to remove solid substances from a liquid or gaseous mixture based on differences in the size of particles

filtering
performing the process of filtration

filter paper
paper with very fine holes (pores) that allow only very small particles to pass through

filtrate
the substance that passes through the filter paper, usually a liquid

residue
what is left in the filter paper after filtration

◀ FIGURE 5.3.2 Filtration produces a filtrate and a residue.

filter funnel
a funnel used to hold filter paper during filtration

Interactive resource
Drag and drop: What is filtration?

In a laboratory, a **filter funnel** is used to hold the filter paper. The funnel may be supported by a funnel ring attached to a retort stand (Figure 5.3.3), or it can sit in the top of a conical flask. The filter paper is carefully folded (Figure 5.3.4) to sit inside the funnel and the mixture is poured into the filter paper. For the filtration to be successful, all of the mixture must go through the paper and not around or through holes in the paper. Therefore, it is important that the:

- mixture does not come to the top of the filter paper
- filter paper is not creased too much
- filter funnel and paper are balanced and not tilted
- filter paper has no tears in it.

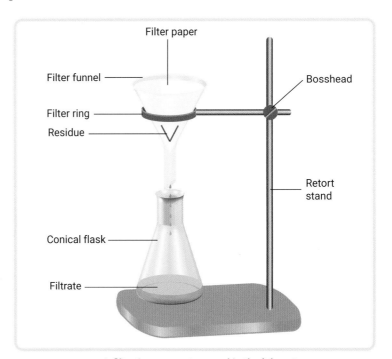

▲ **FIGURE 5.3.3** A filtration apparatus used in the laboratory

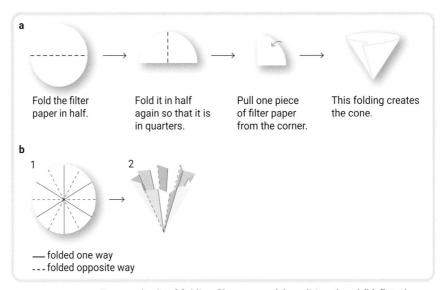

▲ **FIGURE 5.3.4** Two methods of folding filter paper: **(a)** traditional and **(b)** fluted

Separating a suspension by filtration

ACTIVITY 5.3

You need
- mixture of water, calcium carbonate and marbles (approximately 200 mL)
- strainer
- filter paper and filter funnel
- 250 mL conical flask
- 500 mL beaker

What to do
1. Place the strainer above the beaker.
2. Pour the mixture into the strainer.
3. Fold the filter paper in one of the ways shown in Figure 5.3.4 and place it in the filter funnel.
4. Sit the funnel in the top of the conical flask and pour the mixture from the beaker into the funnel.

What do you think?
1. What part of the mixture was collected in the:
   - a strainer?
   - b conical flask?
   - c filter funnel?
2. Why wasn't the calcium carbonate separated from the water by the strainer, but the marbles were?

5.3 LEARNING CHECK

1. **Label** the parts of the filtration set-up shown in Figure 5.3.5.
2. Chloe is filtering a mixture of sand and water.
   - a What will be in the filtrate? **Explain** your answer.
   - b What will be in the residue? **Explain** your answer.
3. After cooking rice in a saucepan, it is strained to remove the water. **State** one similarity and one difference between straining the rice and filtering a suspension using filter paper.
4. **Explain** why filtration will not separate the salt and water in ocean water.
5. Challenge: Ari made a blue copper sulfate solution by adding powdered copper sulfate to water until no more solid could dissolve. He then poured the mixture through filter paper and was surprised to see blue solid copper sulfate as a residue.
   - a Where would Ari have seen the residue?
   - b What colour would the filtrate have been? **Explain** your answer.
   - c **Suggest** why Ari was surprised to see the solid.
   - d **Explain** why there was copper sulfate residue. (Hint: Review Chapter 4.)

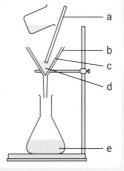

▲ FIGURE 5.3.5 The set-up for filtration

5.4 Separating solutions: evaporation, crystallisation and distillation

BY THE END OF THIS MODULE, YOU WILL BE ABLE TO:
- ✓ define evaporation, crystallisation, distillation, distillate and condenser
- ✓ label the components of a distillation apparatus
- ✓ explain how evaporation, crystallisation and distillation separate mixtures
- ✓ compare evaporation, crystallisation and distillation.

Video activities
Salt: separating mixtures
Fractional distillation

Interactive resource
Label: Lab equipment used to separate mixtures

Extra science investigation
Rate of evaporation of water

GET THINKING

A solar still is a device that uses energy from the Sun to obtain fresh drinking water from dirty water or salt water. Figure 5.4.1 is a diagram of a simple solar still. Use the labels to follow how the solar still works. This same principle is used in the separation techniques in this module.

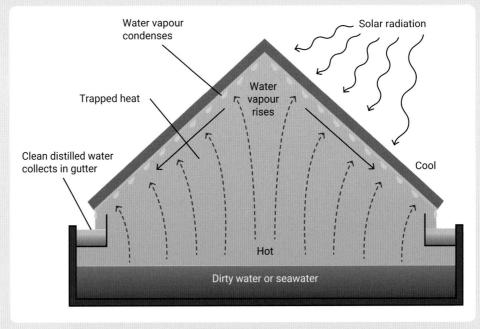

▲ FIGURE 5.4.1 A solar still

Separating solutions

Separating the parts of a solution relies on the different properties of the solute and solvent. When choosing a method, you need to think about which part, or parts, of the solution you want to collect in its pure form.

Evaporation

evaporation
the process of changing from a liquid to a gas at a temperature lower than the boiling point

Evaporation is the process of a liquid changing into a gas. Unlike boiling, evaporation can happen at temperatures lower than the boiling point. For example, water will evaporate when the temperature is 25°C, even though its boiling point is 100°C.

Evaporation can be used to separate the parts of a solution because one part will evaporate more easily than the other. In most cases, the solvent will evaporate to leave only the solute.

After a day at the beach, you will often find salt crystals on your skin. Seawater is a solution of salt and water. When the water evaporates, it leaves the salt behind, which you see as salt crystals. On a hot day, this happens more quickly because the water can evaporate at a faster rate. Around the world, salt evaporation ponds are used to collect salt from seawater.

An **evaporating basin** is used in a laboratory to evaporate the solvent from a solution. The basin is placed on a tripod stand over a Bunsen burner to heat the solution (Figure 5.4.3). Once the water is evaporated, the solute remains in the evaporating basin

▲ FIGURE 5.4.2 A salt lake near Mt Eba in South Australia after the water has evaporated

evaporating basin
a small porcelain dish used to evaporate solvent from a solution

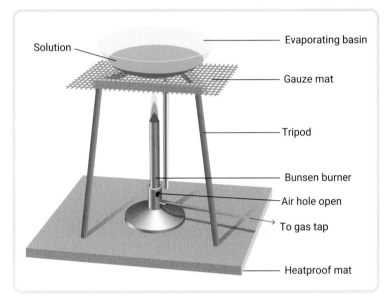

▲ FIGURE 5.4.3 The equipment used to evaporate a solution in the laboratory

crystallisation
the process in which excess solute in a solution forms crystals

Crystallisation

The solubility of a solution is measured by the amount of solute that can dissolve in a certain volume of solvent. Once the maximum amount of solute is dissolved, any additional solute will remain as an undissolved solid.

Crystallisation occurs when some of the solute comes out of solution and forms a solid because it cannot stay dissolved. This happens if the temperature of a solution is lowered, or the volume is reduced when water is removed. Once the crystals have formed, the liquid can be poured off, leaving crystals of the solute. Therefore, the solute has been separated from the solvent.

▲ FIGURE 5.4.4 Crystals of solute forming from a solution as it cools

Distillation

distillation
the process used to separate solutions that collects both the solute and the solvent

Distillation is a process used to separate solutions when we want to keep both the solute and the solvent. The substances in the mixture need to have different boiling points so that they change state between liquid and gas at different temperatures.

The distillation apparatus (see Figure 5.4.5) can separate a solution as it is heated. When a substance in the mixture reaches its boiling point, it forms a gas. The gas rises through a tube and passes into a **condenser**. The condenser has cold water running around its outside, which cools the gas so that it condenses, or forms a liquid. The liquid, called the **distillate**, collects at the end of the tube.

condenser
the piece of equipment in a distillation apparatus that cools the gas so that it changes to liquid

distillate
the liquid collected during the distillation process

Distillation has many applications. Crude oil is a mixture of different compounds, including petrol, diesel and kerosene. Distillation separates the crude oil into its components. Distillation is also used to produce drinkable water from seawater.

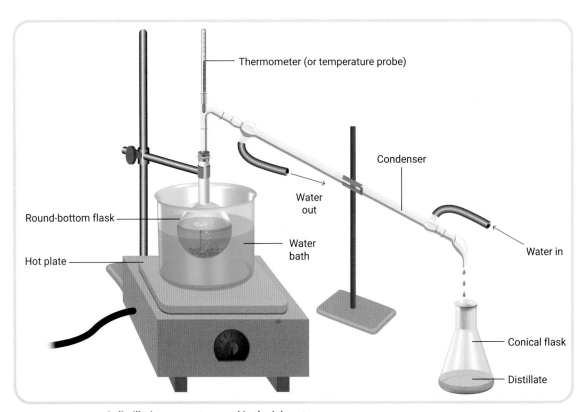

▲ **FIGURE 5.4.5** A distillation apparatus used in the laboratory

Making crystals

✮ ACTIVITY 5.4

Warning
- Hot equipment can cause burns. Do not touch hot equipment. Use tongs or heatproof mitts. Use heatproof mats. Turn off the Bunsen burner after use, and allow all equipment to cool before attempting to move any part of the equipment.
- Chemicals can cause damage if they are inhaled, are ingested or enter your eyes. Always wear safety glasses, lab coat and gloves. Do not inhale or ingest chemicals.
- Broken glass can cut skin. Clean up any broken glass immediately and put it in the glass bin.

Before carrying out an experiment, remember that it is important to read through the instructions and listen to, or read, the safety instructions given by your teacher.

Aim
To obtain pure crystals from a solution

You need
- 10 g potassium aluminium sulfate (common alum) $KAl(SO_4)_2.12H_2O$
- $KAl(SO_4)_2.12H_2O$ seed crystal
- 70 mL distilled water
- electronic balance
- 100 mL measuring cylinder
- two 250 mL beakers
- Bunsen burner
- tripod
- heatproof mat
- gauze mat
- stirring rod
- fine string
- pencil

What to do
1. Weigh 10 g of alum and place it into a clean beaker.
2. Measure 70 mL of distilled water in a measuring cylinder.
3. Add the water to the beaker with the alum.
4. Set up the Bunsen burner, tripod and gauze mat on the heatproof mat.
5. Light the Bunsen burner and turn the flame to a blue flame.
6. Place the beaker on the gauze mat and gently heat the solution.
7. Use the stirring rod to stir the solution until all of the alum has dissolved.
8. Turn off the Bunsen burner and allow the beaker to cool enough to remove it from the tripod.
9. Tie the seed crystal onto the fine string and tie the string onto the pencil at a length so that the seed crystal will sit in the solution in the beaker.
10. Place the pencil on top of the beaker and suspend the seed crystal in the solution.

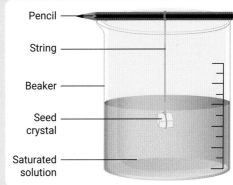

▲ FIGURE 5.4.6   Making crystals

11 Leave the beaker for approximately a week.

12 Each day, observe the beaker and record your observations.

Results

Record your daily observations in a table.

What do you think?

1 Describe the type of solution that you made.
2 Which part of the solution were you able to obtain in a pure form?
3 Explain why the crystals formed from the solution.
4 Suggest a reason for using the seed crystal.

Conclusion

What conclusion can you make regarding the separation of a solution by crystallisation?

Extension

1 Predict what factors may affect the size of the crystal. Plan an investigation to test your prediction.
2 Describe a procedure for recovering as much of the solute as possible.

5.4 LEARNING CHECK

1 **Define**:
   a distillate.
   b condenser.
   c evaporating basin.
2 What physical property is different between the parts of a mixture that are separated by distillation?
3 **Explain** why distillation has an advantage over evaporation for separating a solution.
4 **Describe** what you could do to stop crystallisation happening in a solution.
5 **Predict** which container would result in a liquid evaporating quicker: a test tube or an evaporating basin? Give a reason for your answer.
6 Fractional distillation is a special type of distillation. Conduct research to learn how fractional distillation works and what it is used for.
7 **Compare** distillate and filtrate – how are they similar and how are they different?

5.5 Other separation techniques: magnetic separation, flocculation and chromatography

BY THE END OF THIS MODULE, YOU WILL BE ABLE TO:
- ✓ define magnetism, flocculation and chromatography
- ✓ explain how magnets, flocculation and chromatography separate mixtures.

GET THINKING

Look at the pictures in this module. What do you think they are showing? How do you think magnets, flocculation and chromatography separate mixtures? Annotate a diagram with your predictions. As you work through the module, add to or correct your annotations.

Magnetic separation

One physical property of matter is **magnetism**. This is the force experienced by metals such as iron, steel, nickel and cobalt. Most substances, such as aluminium, plastic, cardboard and wood, are not magnetic and so are not attracted to a magnet.

A **magnet** can remove magnetic substances from a mixture. One application of this separation technique is recycling. Recycling centres use powerful magnets to separate steel cans from aluminium cans. Scrap metal yards use large magnets to separate iron for recycling (Figure 5.5.2).

▲ **FIGURE 5.5.1** A magnet can separate iron nails from sawdust.

magnetism
a force that is experienced by metals such as iron

magnet
a material that produces a magnetic field

Quiz
Separation techniques

Interactive resource
Crossword: Separating substances

Extra science investigations
Does ink contain water?
Flocculation

▲ **FIGURE 5.5.2** Magnetic separation in a scrap metal yard

Chapter 5 | Separating mixtures 139

▲ FIGURE 5.5.3   A colloid before and after the addition of a flocculant

Flocculation

flocculation
the process in which particles in a colloid join to form larger clumps

flocculant
a chemical added to a colloid to make the particles clump together

chromatography
a process used to separate mixtures on the basis of their solubility

The particles in a colloid are so small that they are difficult to remove by methods such as filtration. **Flocculation** is a separation process that involves adding a chemical called a **flocculant** to the colloid. This chemical makes the colloid particles join to form larger clumps. The clumps settle and can then be removed by decanting, filtering or centrifuging.

Flocculation is commonly used to remove chemicals, micro-organisms and solids from water. It is used in swimming pools to remove fine particles so that the water becomes clear. Flocculants are also used to clean water to make it suitable to drink.

Chromatography

Chromatography literally means 'colour writing' and is used to separate mixtures on the basis of their solubility, mass, size and charge. There are many different types of chromatography, such as paper chromatography, thin-layer chromatography and gas chromatography. All types of chromatography use a mobile phase, or solvent, that moves over a stationary phase, which stays in the same place.

Paper chromatography can be used to separate different colours in mixtures, such as food colourings, pigments or dyes. The mixture is carried by a solvent (the mobile phase) along the paper (the stationary phase). Each part moves at a different speed depending on its solubility in the solvent. The more soluble it is, the faster it will travel. In this way, the parts of the mixture become separated from one another, as seen in Figure 5.5.4.

Other more sophisticated forms of chromatography are used in many areas of science. For example, forensic scientists use gas chromatography to test blood and urine samples for the presence of drugs or alcohol.

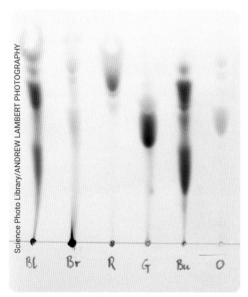

▲ FIGURE 5.5.4   Paper chromatography separating colours from different dyes. Note how different colours have separated at different places along the chromatography paper.

5.5 LEARNING CHECK

1 **Name** the type of separation technique used in the following diagrams.

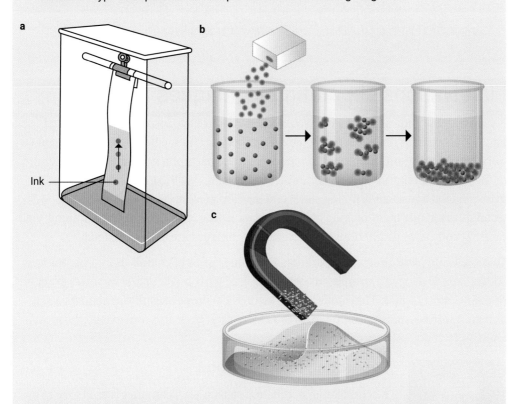

2 Copy and complete the table.

	Magnetic separation	Flocculation	Chromatography
Physical property that differs			
Type of mixture that is separated			
Example			

3 **Explain** how flocculation is used to separate a colloid.

4 Chromatography is used in forensic science. For example, it is used to identify ink in documents.
   a **Describe** how chromatography could be used to identify the ink.
   b **Discuss** whether this should be the only evidence used to convict a suspect.
   c Perform chromatography to identify the components in the ink of your pens. Start by adding water to a cup or mug to a height of 0.5 cm. Next, put a dot of ink about 1 cm from the bottom of a strip of filter paper. If you don't have filter paper, use some sturdy paper towel. Insert the filter paper into the water, with the top of the paper supported by a pencil. Make sure that your dot is not in the water. Watch the water move up the paper. If your ink is water soluble, it will move too. If your ink is not water soluble, try using other solvents but check with your parents or teacher first.

Chapter 5 | Separating mixtures

5.6 Separating techniques traditionally used by First Nations Australians

FIRST NATIONS SCIENCE CONTEXTS

IN THIS MODULE YOU WILL:
✓ examine First Nations Australians' use of different separating techniques.

First Nations separating techniques

For many thousands of years, First Nations Australians have selected, processed and used natural material for many purposes, including cooking and medicine. They have applied a variety of separation techniques to separate out the parts of a mixture that are useful or the parts of a mixture that are harmful. This has allowed First Nations Australians to obtain food and clean drinking water and develop medicines. These separation techniques include hand-picking, winnowing, yandying, sieving, filtering, straining, cold-pressing and steam distillation. Many of these techniques are still in use today.

To successfully perform these separation techniques prior to colonisation, First Nations Australians developed a complex understanding of what made up the mixtures available in their Country/Place. They understood the physical properties of the mixtures, such as their hardness, density, boiling point and solubility. They also understood the most effective techniques for each mixture's separation.

Separating to obtain food

▲ FIGURE 5.6.1 An example of a *coolamon*, origin unknown

A combination of different separation techniques is used to obtain seeds for cooking and eating. For example, the Alyawarre Peoples of the Sandover River region in the Northern Territory harvested about 36 different seed types for food. After collecting the seed pods, they beat them with sticks (a process known as threshing) to break the pods and release the seeds. A mixture containing seeds, broken pods, sticks and dirt was then scooped up from the ground using a specially made wooden bowl, called an *arlengarr*. Different First Nations cultural groups had a specific name and, at times, design for these wooden dishes. Since colonisation, the term *coolamon* has been applied to containers of the type seen in Figure 5.6.1.

Large pieces of unwanted material are also removed from the mixture in the *coolamon* by hand-picking. The remaining mixture is then winnowed (Figure 5.6.2). In this process, the mixture is thrown in the air to allow the lighter particles to be blown away and the heavier particles, such as seeds, to be caught in the *coolamon*. This process must be repeated many times to remove all the lighter particles and leave behind only the seeds.

Yandying, like winnowing, is used to separate components of a mixture with different densities, such as seeds and sand. This process is similar to how people pan for gold. In this process, the mixture is placed in a wooden container. One corner of the container is raised, and the container is gently shaken back and forth, causing denser particles to move to the bottom while less dense particles remain at the top.

Mulga seeds were an important food source for the Yankunytjatjara Peoples of north-west South Australia. Before mulga seeds could be consumed, they had to undergo two yandying processes. The collected seed pods would first be threshed and rubbed to release the seeds. The resulting mixture was then yandied to separate the seeds from any remaining pod pieces. The clean seeds were then baked in hot sand and ashes. The mixture of seed, sand and ashes was then winnowed and yandied to again separate the seeds.

▲ FIGURE 5.6.2 Winnowing to separate a mixture of seeds

Special woven bicornate-shaped ('bicornate' means two-horned) baskets made by the Bama Peoples of tropical north Queensland have long been used as sieves, strainers or filters. The Dyirbal People call them *jawun* (shown in Figure 5.6.3). Baskets may be loosely woven with larger holes for washing away sand and dirt from food or straining water from food. Baskets may also be tightly woven with small holes. These were used to strain water for drinking. They could also be filled with foods that needed to have toxins removed before eating, such as the cooked, thinly sliced seeds of a cycad, which were then placed in running water to leach away toxins.

Separating to obtain clean water

Sand and various types of plant parts were also used to filter water for drinking. For example, in traditional times the Gunditjmara Peoples of south-west Victoria used flowering honeysuckle (banksia) cones to filter dirty water for drinking. The cone would be used like a straw. Sucking water through the cone would exclude impurities so that the water would be clean enough to drink.

▲ FIGURE 5.6.3 A bicornate-shaped basket, known as *jawun*, used by the rainforest peoples of north-eastern Queensland to strain, sieve and filter mixtures

First Nations Australians developed effective filters from the stems and hollow stalks of many different plants, such as the lotus lily, bamboo, bulrush and swamp panic grass. These were used as drinking straws to screen out impurities and provide clean water for drinking.

▲ FIGURE 5.6.4 The Gunditjmara Peoples used banksia cones to filter drinking water.

ACTIVITY 1

1. **Design** a method to separate rice from a mixture containing leaves, rice, grass clippings and sand.
   a. **List** the steps needed to separate each component. **Name** the separation technique used in each step.
   b. For each step **identify** the properties used to achieve the separation of each specific component.
2. **Explain** what properties are used to remove toxins from food by leaching.
3. **Propose** how the traditional knowledge of First Nations Australians of using plants for drinking straws could be applied to reduce the environmental impact associated with the use of plastic drinking straws.

Separating to obtain medicines

For thousands of years, First Nations Australians have understood, extracted and used the medicinal components of native plants. Traditionally, a variety of plant oils would be extracted by steam distillation. By placing fresh, wet plant matter over cool fires, steam was used to release the medicinal oils in the leaves which could then be inhaled. For example, the Bundjalung Peoples of northern New South Wales inhaled vapours from the heated leaves of tea tree to effectively treat the symptoms of coughs and colds.

The medicinal component of plant material can be obtained by boiling or soaking the plant material in hot or cold water, resulting in a solution that contains the desired component. This mixture can then be drunk or applied to wounds. For example, the oil from eucalyptus leaves was widely used by many First Nations Australians in steaming, in a drink and as a wash to treat wounds. The Pitjantjatjara Peoples of north-west South Australia make a drink to relieve sore throats by soaking the foliage of the flat sedge plant in hot water.

Cold-pressing is another traditional technique used to extract medicinal components from plant material. One advantage of this technique is that no heating is required. To cold-press, the plant material is ground to a pulp and then pressed to obtain oils, which can be applied externally.

▲ **FIGURE 5.6.5** Tea tree leaves and flowers

ACTIVITY 2

1. a. **Identify** three separation techniques used to obtain medicines.
   b. For each technique identified, **explain** what properties are used in the separation process.
2. **Discuss** an example of where one of these techniques and medicines is used today.

5.7 Separation techniques used in recycling

SCIENCE AS A HUMAN ENDEAVOUR

BY THE END OF THIS MODULE, YOU WILL BE ABLE TO:
- explain how separation techniques are used in recycling.

People are becoming more aware of the importance of recycling. During recycling, the materials are broken down into their raw materials and then made into new products. For example, one company, Precious Plastic in Margaret River, Western Australia, melts plastic lids and then creates products such as plant pots.

Different methods of recycling are used for different materials. Therefore, the items must be separated according to what they are made from. You can start this at home by:
- separating plastic bottles, glass bottles and aluminium cans before taking them to Containers for Change
- taking scrap metal to a scrap metal yard
- taking batteries, light globes, soft plastics and mobile phones to recycling centres
- using green waste (plant material) to make compost
- putting paper recyclable materials into the recycling bin.

▲ FIGURE 5.7.1 A plant pot made from recycled plastic lids

The materials in your recycling bin are separated at the recycling centre. The following steps are involved in separating the wastes.

1. The waste passes over a turning cylinder with holes. Heavy items, such as plastic, glass and metal, fall through, leaving paper and cardboard.
2. Workers remove cardboard from paper as it moves past them on a conveyor belt.
3. The heavy items move along a separate conveyor belt. As they pass under a magnet, the steel cans are attracted to it and lifted away from the other materials.
4. Workers manually separate milk bottles, aluminium cans and plastic bottles as they move along a conveyor belt.

Video activity
How recycling works

▲ FIGURE 5.7.2 A magnet separates the steel from other recyclable materials.

5.7 LEARNING CHECK

1. **List** three physical properties used in the separation of recyclable materials.
2. A turning cylinder separates the heavy products from the lighter products. **Name** the separation technique utilised in this step.
3. **Explain** why manual sorting is needed to separate paper and cardboard.
4. A new technology for separating plastics uses invisible barcodes that allow a scanner to identify the type of plastic and whether it can be recycled. **Discuss** the value of this technology, identifying an advantage and a disadvantage of adding the barcode to products.

SCIENCE INVESTIGATIONS

5.8 Writing a method

SCIENCE SKILLS IN FOCUS

IN THIS MODULE, YOU WILL FOCUS ON LEARNING AND IMPROVING THESE SKILLS:

▶ separating a mixture based on differences in physical properties and by applying an understanding of separation techniques.

In science, a method is what you did to collect your data during an investigation. It is slightly different from a procedure, which is the steps that you are planning to do. During an investigation, it is common for the steps to be changed. Therefore, the method is a more accurate representation of what was done.

When writing your method, it should be:
- written in numbered steps to show the order that you did things
- clear and easy to understand
- detailed so that others can repeat what you did
- written in the past tense because it is something that you did in the past
- written in the passive voice so that the objects or equipment are the focus, not the people.

An example of a step from a method is:

1 The water was poured into the clean beaker.

Video
Science skills in a minute: Methods

Science skills resource
Science skills in practice: Writing a method

SEPARATING A MIXTURE

AIM

To separate a mixture

YOU NEED

- ☑ mixture supplied by your teacher consisting of solid calcium carbonate (insoluble) and solid copper sulfate (soluble). Your teacher will know exactly how much of each chemical is in the mixture
- ☑ 250 mL beaker
- ☑ stirring rod
- ☑ distilled water
- ☑ equipment for a range of separation techniques, e.g. filter paper, filter funnel, conical flask, Bunsen burner, evaporating basin

 Warning
- Hot equipment can cause burns. Do not touch hot equipment. Use tongs and heatproof mats.
- Chemicals can cause damage if they are inhaled, are ingested or enter your eyes. Always wear safety glasses, lab coat and gloves. Do not inhale or ingest chemicals.

WHAT TO DO

1. Work in groups of two or three to develop a plan for separating the mixture.
2. Write a procedure or draw a flow chart showing your planned steps.
3. Show your procedure to your teacher.
4. Use your procedure to separate your mixture. You may need a couple of days to complete this.
5. Take photos of the steps that you used. Add annotations to label key equipment. If you cannot take photos, draw a scientific diagram with pencil and ruled lines.
6. Make a note of any changes to your procedures or any details that you missed.
7. Use an electronic balance to measure the mass of each part of the mixture. Record this in your results table (Table 5.8.1).
8. Ask your teacher how much of each chemical was originally added to the mixture. Record this in your results table.

WHAT DO YOU THINK?

1. Use your procedure, photos and notes about any changes to write a detailed method of what you did to separate the mixture.
2. For each technique you used, write a paragraph explaining its relevance. Include the physical property that it related to and what happened to each part of the mixture.
3. Compare the masses that you collected with the masses that your teacher originally added. Suggest a reason for any differences.

CONCLUSION

Write a short paragraph summarising how your method separated the mixture.

RESULTS

Copy and complete Table 5.8.1.

▼ TABLE 5.8.1 Masses of parts of a mixture

Part of the mixture collected	Mass collected after separation (g)	Mass originally added to the solution (g)
Calcium carbonate		
Copper sulfate		

5 REVIEW

REMEMBERING

1. **Match** each piece of equipment with its use.

Equipment	Use
Condenser	Separates a solid sediment from water very quickly by spinning
Centrifuge	Separates substances according to the size of the particles
Filter paper	Cools vapours and turns them back into a liquid
Evaporating basin	Separates solutions into their parts by boiling
Magnet	Holds filter paper during filtration
Distillation apparatus	Separates iron, cobalt and nickel from other substances
Funnel	Evaporates water from a solution while it is gently heated

2. **State** whether each of the following separation techniques is used to separate a suspension, colloid or solution.
   a. Distillation
   b. Flocculation
   c. Centrifuging
   d. Crystallisation

3. **State** the name given to the substance:
   a. left in the filter paper.
   b. collected at the end of distillation.
   c. that passes through a filter.
   d. added to a colloid to make the particles clump together.

4. **Draw** a labelled scientific diagram of the equipment you would use in the laboratory to recover the salt from a salt solution.

5. **Name** the parts (a–h) of the distillation equipment in the following diagram.

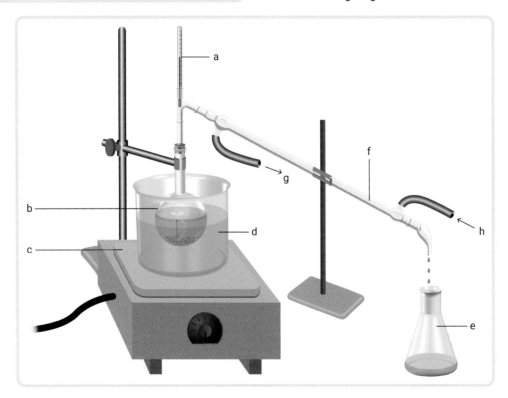

6 **Describe** the function of the following parts of distillation equipment.
   a Thermometer
   b Condenser
   c Hot plate or Bunsen burner

UNDERSTANDING

7 **Explain** why distillation is used instead of evaporation to separate some solutions.
8 **Identify** which technique(s) could be used to:
   a collect salt from seawater.
   b determine the colours in black jelly beans.
   c remove large pebbles from a mixture of soil, leaves and dirt.
   d collect drinking water from seawater.
   e determine if an Olympic sprinter has any illegal chemicals in their blood.
9 **List** the methods of decanting, filtration and sedimentation in order from most to least effective for separating a suspension of sandy water. **Explain** your answer.
10 **State** whether the following statements are true or false. For those that are false, rewrite them so that they are true.
   a Only solutions can evaporate.
   b If a salt solution is left on the windowsill for long enough, the solute will evaporate, leaving the solvent behind.
   c Increasing the temperature of a solution will increase the solubility of the solute in the solvent.
   d Crystallisation can occur only by cooling a saturated solution.
11 **Describe** the similarities and differences between the processes of evaporation and crystallisation.
12 **Explain** how flocculation can make the water in a spa clear.
13 **Explain** how distillation is similar to water evaporating from the ocean and then falling as rain.

APPLYING

14 First Nations Australians use yandying and winnowing to separate mixtures. **Compare** these two techniques, **identifying** the circumstances when each is used.
15 **Describe** how sedimentation and decantation are used in the kitchen to separate peas from the water they were cooked in.
16 **Compare** the filter paper used in the laboratory with a strainer used in the kitchen.
17 Baleen whales are called filter feeders because they take huge amounts of water into their mouths to separate out fish and krill for food, and expel the water. They are called filter feeders. **Suggest** what the structure of a baleen whale's mouth looks like. **Explain** your answer.
18 The Dead Sea receives very little rainfall. The air is dry and temperatures are always high. Use this information, and your understanding of saturated solutions, to **explain** why the Dead Sea coastline is covered in solid salt.
19 When using chromatography to separate a mixture, only small amounts of the mixture are required. Why would this be helpful if gas chromatography was used to determine the percentage of alcohol in a blood sample?
20 In dairies, fresh milk is separated into cream and milk. Originally this was done by leaving the fresh milk until the cream rose to form a layer on top that could be skimmed off. In 1894, Gustaf de Laval patented the first milk separator. The fresh milk was placed in the separator. When the handle was turned, the separator bowl would spin at thousands of revolutions per second.
   a What type of mixture is fresh milk?
   b **Outline** the properties of the cream and milk that are used in each of these separation techniques.
   c What type of separation technique was used initially? **Describe** this process.
   d What type of separation technique was used by the milk separator? **Explain** how this was able to separate the cream from the milk.
21 You are camping in the outback, and have run out of water. The water in the nearby waterhole looks rather murky but you have seen animals drinking it. Using anything you brought along with you, what steps would you take to clean this water so that it is likely to be safe to drink? You can use drawings to explain what you will do.
22 What are some methods that you have used at home without realising they were physical separation techniques?

EVALUATING

23 A mixture of sand, salt and water to be separated in the laboratory weighed 235.6 g. After the mixture was filtered, the residue was wet and the filtrate weighed 202.9 g. The filtrate was then distilled until 4.03 g of white powder was left in the flask.
   a **Identify** the residue, filtrate and distillate.
   b From the data given, **determine** the mass in the original mixture of the:
      i sand.
      ii salt.
      iii water.
   c After filtration, the residue was still wet. **Assess** the accuracy of the masses obtained from this activity.
   d How could this experiment be done in a different way to improve the accuracy of the mass measurements of the sand, salt and water? Use a flow chart to show your modified method.

24 There was a break-in at a jewellery store. In the store, there is a note written in black ink. Police question three people: person 1, the store owner (who came from home after being notified); person 2, a young woman (who reported hearing the alarm and notified police); and person 3, a local security guard. Each person had a different black ink pen. Forensic scientists use chromatography to compare the ink in each person's pen with the ink on the note. The results are shown below.

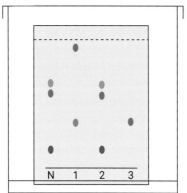

   a Whose pen was used to write the note? **Explain** how you determined this.
   b Does this mean that person was the thief? Give reasons for your answer.

25 In 2010, two students from Rice University in Texas, United States, created a cheap centrifuge that could be used to separate blood without using electricity. They made their centrifuge from a salad spinner at a cost of about $30.
   a What features of a salad spinner would make it suitable for separating blood?
   b What modifications would be needed, and why?
   c What impact could this invention have in underdeveloped countries?

26 Another separation technique uses a separating funnel, as shown in the diagram below.

Use the information provided and your knowledge of physical properties to **explain** how a separating funnel separates a mixture of oil and water.

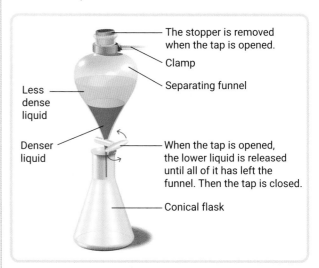

CREATING

27 **Create** a concept map to organise the information you have learned about the different separation techniques.

28 Salt is only slightly soluble in a liquid called ethanol. Ethanol has a boiling point of approximately 78°C. Write a detailed method you could use in the laboratory to separate a mixture of sand, gravel, ethanol and salt. The sand, gravel and salt must be reclaimed as solids, and you must also reclaim the liquid ethanol. **Explain** how each separation process allows the collection of each substance.

BIG SCIENCE CHALLENGE PROJECT #5

Your challenge is to separate a mixture of sugar, flour and staples so that you have each of the substances in their original, solid form.

1 Connect what you've learned

In this chapter, you have learned about physical properties and methods to separate mixtures.

a Use a table to summarise the key physical properties of sugar, flour and staples.

b List the separation methods that are useful to separate substances with these properties.

2 Check your thinking

Use a flow chart to plan how to separate the sugar, flour and staples. Include any equipment that you will need to use.

3 Make an action plan

Check your plan with your teacher – have you remembered everything? Your teacher may let you test your plan to see if it works.

4 Communicate

Create a short video to share how you would separate this mixture.

- If you were able to test your plan, use photos or videos of your experiment to make your video. Add text and/or voice-over to explain what you did and why.

- If you were unable to test your plan, create a series of drawings, images or animations to represent your plan. Add text and/or voice-over to explain what you would do and why.

- If you do not have access to a device to make a video, create a story board to plan what would go into a video. Add labels and notes to your story board cells to show the information that you would share.

Mark Fergus Photography

6 Classifying living things

6.1 Classification (p. 154)
Classification is organising objects into groups based on similarities.

6.2 Dichotomous keys (p. 156)
A dichotomous key sequentially divides groups into two to identify an object.

6.3 Different species (p. 159)
Species are organisms with similar characteristics that produce fertile offspring.

6.4 Linnaean classification of living things (p. 162)
Linnaean classification organises living things into groups that progressively become more specialised.

6.5 The kingdoms of living things (p. 167)
All living things belong to one of five kingdoms.

6.6 Naming living things (p. 171)
Living things are named by their genus and species.

6.7 FIRST NATIONS SCIENCE CONTEXTS: First Nations Australians' classification systems (p. 175)
First Nations Australians have developed classification systems for plants and animals.

6.8 SCIENCE AS A HUMAN ENDEAVOUR: Changing classifications (p. 177)
Classifications may change as new information becomes available.

6.9 SCIENCE INVESTIGATIONS: Creating a dichotomous key (p. 179)
Use the properties of scientific equipment to create a dichotomous key.

BIG SCIENCE CHALLENGE #6

▲ FIGURE 6.0.1 Observe these organisms. How can they be organised?

▶ What do the organisms in Figure 6.0.1 have in common?
▶ What are the main characteristics of each?
▶ How can they be organised into groups?
▶ How are they named?

#6 SCIENCE CHALLENGE ACCEPTED!

At the end of this chapter, you can complete Big Science Challenge Project #6. You can use the information you learn in this chapter to complete the project.

Assessments
- Prior knowledge quiz
- Chapter review questions
- End-of-chapter test
- Portfolio assessment task: Research project

Videos
- Science skills in a minute: Dichotomous keys **(6.9)**
- Video activities: Dichotomous keys **(6.2)**; Carl Linnaeus' classification system **(6.4)**; Naming species **(6.6)**; What is taxonomy? **(6.8)**

Science skills resources
- Science skills in practice: Creating a dichotomous key **(6.9)**
- Extra science investigations: Identifying insects using a dichotomous key **(6.6)**; Lolly dichotomous key **(6.6)**

Interactive resources
- Drag and drop: Get classifying! (6.1); What is a species? **(6.3)**
- Label: Levels of classification **(6.4)**
- Match: Kingdoms **(6.5)**

Nelson MindTap

To access these resources and many more, visit:
cengage.com.au/nelsonmindtap

Chapter 6 | Classifying living things

6.1 Classification

BY THE END OF THIS MODULE, YOU WILL BE ABLE TO:
- ✓ define classification and characteristic
- ✓ describe examples of classification
- ✓ classify objects based on similar characteristics
- ✓ explain the reason for classifying objects.

Interactive resource
Drag and drop: Get classifying!

classification
grouping things according to how similar they are

characteristic
a quality or feature that makes something recognisable

GET THINKING

When you go grocery shopping with your parents, where would you find the following?

- Ice-cream
- Bread
- Washing powder
- Apples
- Frozen pizza
- Cheese
- Wraps
- Dog food
- Yoghurt
- Tomatoes
- Cleaning products
- Cat food
- Bananas

Reflect on your answers. Why is each of the items found in that location? What would happen if they were somewhere else? What other places are organised in a certain way? Why do they do this?

▲ **FIGURE 6.1.1** Zoos have a nocturnal house for animals that are active at night, such as this western ringtail possum at Perth Zoo.

Understanding classification

Classification is the process of organising things into groups based on similar **characteristics**. Organising things into groups based on their similarities provides information about members of the group. For example, at school you are organised into year groups based on age. This means that your teachers know what you have already learned and what you need to learn during the year. In sporting teams, you may be organised into different divisions based on previous successes. This helps your coach know the ability of the team and means that you are playing against other teams of a similar ability. In a zoo, the animals are organised according to their native environments. This helps their keepers provide a suitable habitat and allows visitors to understand where the animals came from.

The effectiveness of a classification system depends on the characteristics on which it is based. The characteristic should be:
- observable so that it can be identified
- consistent so that the object will always be classified in the same way
- related to the reason for classification so that the information is valid.

There are many reasons for classifying things, and the reason will determine the way that they are classified. For example, food may be classified as:
- savoury or sweet
- containing gluten or gluten-free
- eaten when hot, cold or at room temperature
- fresh or processed
- cereal, dairy, fruit and vegetable or meat.

▲ FIGURE 6.1.2 One way of classifying food is as savoury or sweet.

6.1 LEARNING CHECK

1. **Define** 'classification'.
2. What characteristics would you use to help a person identify vanilla icecream who has never eaten it? Choose all options that apply.
   a It is sweet.
   b It is served on a plate.
   c It is eaten cold.
   d It smells and tastes like vanilla.
3. **Describe** how the cutlery (knives, forks and spoons) are organised in your kitchen drawer. **Explain** why they are arranged this way.
4. Use the photo in Figure 6.1.3 to **classify** buttons. **State** the characteristics of the buttons that you would use to classify the buttons into:
   a two groups.
   b more than two groups.
5. **Discuss** the advantages of classifying car drivers into learners (L plates), provisional (P plates) and open (full) licence.

▲ FIGURE 6.1.3 How would you classify the buttons in this photo?

6.2 Dichotomous keys

BY THE END OF THIS MODULE, YOU WILL BE ABLE TO:
- ✓ define dichotomous key
- ✓ apply a dichotomous key to identify an object
- ✓ construct simple dichotomous keys.

Video activity
Dichotomous keys

GET THINKING

Like many scientific terms, the word 'dichotomous' comes from Latin and Greek root words. Use the information in Table 6.2.1 to predict what dichotomous means.

▼ TABLE 6.2.1 The meaning of the word 'dichotomous'

Root word	Language	Meaning/use
di-	Greek	two
dicho-	Greek	in two parts
tom-, tomia	Greek	to cut
-ous	Latin	a suffix that makes a word an adjective (describes a noun)

What is a dichotomous key?

dichotomous key
a tool used by scientists to classify living things; two choices are given at each level, which helps to narrow down what species something is

A **dichotomous key** is a tool used to classify objects based on dividing a group into two new groups. Each new group is then further divided into two groups. This process continues until there is only one object in the group.

Figure 6.2.1 shows a simple dichotomous key for items used to eat with. The original group included plates, bowls, knives, spoons and forks. These were then divided into two groups based on whether they were cutlery (knives, spoons and forks) or crockery (plates and bowls). Each group was then further divided into two groups based on characteristics. Note how the name of the item is listed when it is the only thing in that group.

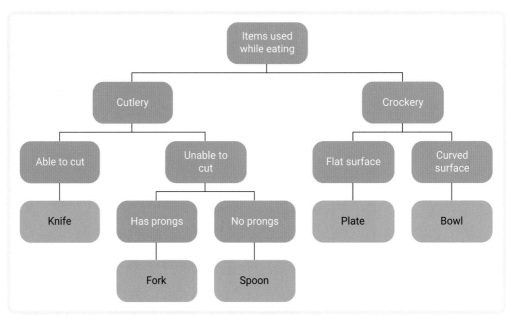

▲ FIGURE 6.2.1 A dichotomous key for items used to eat with

As with any classification system, the effectiveness of a dichotomous key relies on the chosen characteristics. The characteristics for each division should:

- remain the same for that object so that it can always be placed in the same group
- have two options so that all items can be placed in one of the two groups
- be **objective** (not **subjective**) so that everyone will place the item in the same group.

Dichotomous keys are represented in different formats. The most common forms are tree diagrams and linked keys.

objective
not influenced by personal feelings or opinions

subjective
based on personal feelings or opinions

tree diagram
a diagrammatical dichotomous key made by branching, which represents the splitting of each group

Tree diagram

A **tree diagram** (or branching tree) shows the dichotomous key in a visual format that has lines representing each division. Figure 6.2.3 is a tree diagram of a dichotomous key for different balls used in sports.

▲ **FIGURE 6.2.2** Balls used in sports

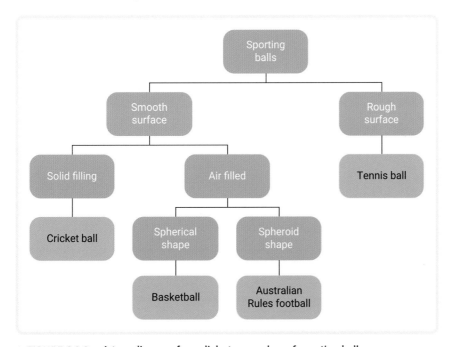

▲ **FIGURE 6.2.3** A tree diagram for a dichotomous key of sporting balls

Linked key

A **linked key** (or **tabular key**) is a descriptive key. It is a numbered list of questions, or statements, with two options. The answer to each question directs you to the next question. Figure 6.2.4 (on the next page) is a linked key for the dichotomous key for sporting balls.

linked key (tabular key)
a descriptive dichotomous key made of numbered questions or statements

Chapter 6 | Classifying living things

1a. Ball has a smooth surface .... Go to 2
1b. Ball has a rough surface .... Tennis ball
2a. Ball is solid .... Cricket ball
2b. Ball is filled with air .... Go to 3
3a. Ball is spherical .... Basketball
3b. Ball is spheroid shaped .... Australian Rules football

▲ FIGURE 6.2.4 A linked key for sporting balls

6.2 LEARNING CHECK

1 What is a feature of all dichotomous keys?
2 Figure 6.2.5 shows a dichotomous key for fruit.

1a. Round in shape .... Go to 2
1b. Not round in shape .... Go to 3
2a. Orange in colour .... Go to 4
2b. Not orange in colour .... Go to 5
3a. Red in colour .... Fruit A
3b. Yellow in colour .... Fruit B
4a. Easy to peel .... Fruit C
4b. Not easy to peel .... Fruit D
5a. Crunchy when eaten .... Fruit E
5b. Soft when eaten .... Fruit F

▲ FIGURE 6.2.5 A dichotomous key for fruit

a What letter in the dichotomous key represents each type of fruit listed in the table.

Fruit	Letter
Red apple	
Banana	
Orange	
Plum	
Mandarin	
Strawberry	

b What type of dichotomous key is Figure 6.2.5?
c **Discuss** whether this dichotomous key would be appropriate to identify an apricot.
d **Discuss** how objective each statement is.
3 **Create** a dichotomous key to identify six items in your pencil case. Represent the dichotomous key in both a tree diagram and a linked key.

6.3 Different species

BY THE END OF THIS MODULE, YOU WILL BE ABLE TO:
- ✓ define species
- ✓ classify organisms as the same or different species.

GET THINKING

There is great diversity among humans, as shown in Figure 6.3.1, yet we are all the same species. What makes us all human? What do you think being the same species means? Discuss this with other students in your group and come up with a definition of species. As you work through this module, reflect on your definition and refine it with your new knowledge.

▲ FIGURE 6.3.1   Humans are a diverse species.

Organisms

In Year 5, you learned that living things can move, respire, respond to stimuli, grow, reproduce, excrete wastes and gain nutrition. All living things are called **organisms**.

organism
a living thing

Interactive resource
Drag and drop: What is a species?

▲ FIGURE 6.3.2   The koala is gaining nutrition from the leaves of the eucalyptus tree. The koala and eucalyptus are both organisms.

Species

species
a group of similar organisms that can breed to produce fertile offspring

fertile
able to produce offspring

Organisms are commonly classified into groups called **species**. All members of a species have similar characteristics and can breed with one another to produce **fertile** offspring, which means that the offspring can also produce offspring.

Sometimes it is easy to classify organisms into their species. For example, you would easily identify an organism as a dog or a cat or a horse. Other organisms may be more difficult to tell apart. For example, a Damara sheep looks very similar to a goat (Figure 6.3.3). In these instances, scientists use information other than an organism's looks to classify them.

▲ **FIGURE 6.3.3** (a) The Damara sheep and (b) goat. Athough they look similar, they cannot interbreed to produce fertile offspring, so they are classified as different species.

Hybrid

sterile
cannot produce offspring

hybrid
the offspring of a mating between two different species

Some species can interbreed, but the offspring are **sterile**. This means that they cannot reproduce. These offspring are called **hybrids**.

An example of this is the donkey and horse. If a male donkey breeds with a female horse, they produce a mule. Mules are sterile and therefore are a hybrid and not a species.

▲ **FIGURE 6.3.4** A mule is a hybrid produced by a cross between two species: a donkey and a horse.

Classifying within species

You just need to look at a group of dogs to see that there is a lot of variation within a species. When organisms are adapted to live with humans, they are **domesticated**. By controlling the breeding of domestic organisms, humans have produced organisms with specific features. These are known as different **breeds**. For example, Great Danes and chihuahuas are both the same species but are different breeds.

domesticated
adapted over time to live with humans

breed
a group of organisms of the same species with distinctive features

▲ **FIGURE 6.3.5** Great Danes and chihuahuas are both in the same species but are different breeds of dogs.

6.3 LEARNING CHECK

1 **Define**:
   a species.
   b hybrid.
   c breed.
2 **Describe** the difference between fertile and sterile.
3 Humans look very different from one another. **Explain** why humans are all classified as the same species.
4 **Explain** why it is advantageous to classify organisms into species.
5 A liger is produced from a male lion and a female tiger (Figure 6.3.6). **Discuss** whether a liger is a species or a hybrid.

▲ **FIGURE 6.3.6** A liger

6.4 Linnaean classification of living things

BY THE END OF THIS MODULE, YOU WILL BE ABLE TO:
- ✓ describe the Linnaean classification of living things
- ✓ discuss the advantages of a common method of classifying living things
- ✓ justify the size of each group in the Linnaean classification system.

Video activity
Carl Linnaeus' classification system

Interactive resource
Label: Levels of classification

> **GET THINKING**
>
> Mnemonics can help you remember lists of words or facts. In this module, you will need to remember these words, in order:
>
> **K**ingdom, **P**hylum, **C**lass, **O**rder, **F**amily, **G**enus and **S**pecies.
>
> Use the first letter of every word to make a saying that you will remember. For example, **K**ings **P**refer **C**rowns **OF G**reat **S**ize. Use this during the module to remember the words.

Linnaean classification system

Linnaean classification system
a classification system consisting of a hierarchy of groups, with each group being further divided into smaller groups based on similar characteristics

taxonomy
the study of naming, defining and classifying organisms

taxa
groups in the classification system of organisms (singular: taxon)

There is a huge number of different organisms on Earth. One estimate is that there are 8 700 000 different species in total, 6 500 000 of them living on land and 2 200 000 living in the oceans. With more than 8 million organisms, an efficient method of classifying them is important. The **Linnaean classification system** used by scientists provides information about the characteristics of organisms in a group and the relationship between organisms. This system is used internationally, providing a common method of classification for scientists everywhere. This allows them to:

- communicate with clarity
- understand each other's work
- collaborate to build an understanding of living things.

The science of classifying organisms is called **taxonomy** because it involves organising organisms into groups called **taxa**.

▲ FIGURE 6.4.1 A statue of Carl Linnaeus – the father of taxonomy

Carl Linnaeus

Carl Linnaeus is considered the father of taxonomy. He was born Carl von Linné on 23 May 1707, at Stenbrohult, in southern Sweden. His father, a Lutheran pastor, was a keen gardener and so young von Linné loved plants from when he was young. He disappointed his parents by showing no interest in the priesthood; instead, he entered the University of Lund in 1727 to study medicine. A year later, he transferred to the University of Uppsala, the most prestigious university in Sweden. He spent most of his time at Uppsala collecting and studying plants. Despite being poor, he organised expeditions to Lapland in 1731 and to central Sweden in 1734 to study plants.

Carl von Linné's work led him to develop a classification system of living things that was published in 1735 in the *Systema Naturae*. In 1741, he gained a professorship at Uppsala. Here he restored the university's garden, arranging the plants according to his system of classification. Carl von Linné Latinised his name and became known as Carolus Linnaeus.

Divisions in the Linnaean system

The Linnaean classification system groups organisms based on their **structure** and characteristics. Because these features are inherited, the groupings indicate the degree to which organisms are related. All living things are initially classified into large groups based on a few similarities. Each group is then divided into smaller and smaller groups based on more and more similarities. This continues until there is only one species in each group.

structure
how something is built or organised

The first level of grouping is kingdom. There are five kingdoms: Animalia, Plantae, Fungi, Protista and Bacteria. Some scientists also include a sixth kingdom of Archaebacteria. You will learn more about the different kingdoms in Module 6.5.

Each kingdom is divided into smaller groups called phyla. Each phylum divides into classes, each class divides into orders, each order divides into families, each family divides into genera (singular: genus) and each genus divides into species. This division is summarised in Figure 6.4.2.

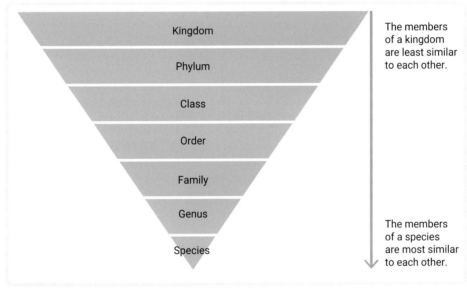

▲ FIGURE 6.4.2 The levels of Linnaean classification

As you move down the **hierarchical** levels, there are fewer organisms in each group. However, the organisms become more and more similar because, at each division, the characteristics used to organise the groups become more specific. This increased similarity is due to organisms being more closely related. Therefore, scientists use taxonomy to indicate how closely species are related.

hierarchical
an order of importance

A deck of cards can be used as an **analogy** to explain this. In the whole deck, there are 52 cards. If we split it into red cards and black cards, there are now only 26 cards in each group. Cards within each group have two characteristics in common – they are cards and they all have the same colour. If we split the red cards into hearts and diamonds, there are now only 13 cards in each group. However, they now have three things in common – they are a card, they are the same colour, and they have the same symbol.

analogy
a comparison

Figure 6.4.3 outlines the classification of humans and Table 6.4.1 outlines the classification of the river red gum. Note how each species is classified into a particular group at each hierarchical level. Also note that plant classification uses division instead of phylum.

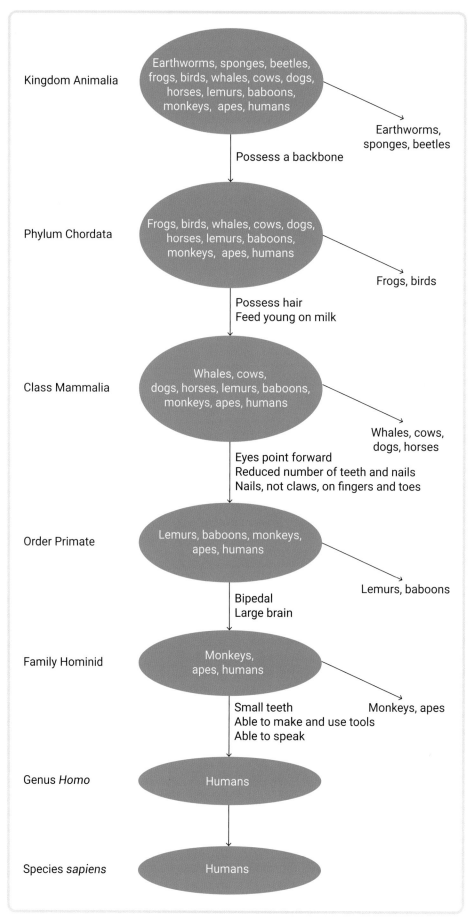

▲ FIGURE 6.4.3 The classification of humans

▼ TABLE 6.4.1 The classification of river red gums

Level of classification	Example	Features of the group	Organisms found in this group
Kingdom	Plantae	• Multicellular • Complex cell structure, produce their own food by photosynthesis	Eucalypt, pine tree, pomegranate, guava, grass, fern, moss, green algae
Division*	Magnoliophyta	• Possess a transport system • Have seeds and flowers	Eucalypt, pomegranate, guava
Class	Magnoliopsida	• Dicotyledons – have two seed leaves	Eucalypt, pomegranate, guava
Order	Myrtales	• Woody plants, often with flaky bark	Eucalypt, pomegranate, guava
Family	Myrtaceae	• Evergreen leaves • Leaves contain essential oils	Eucalypt, guava
Genus	*Eucalyptus*	• Gumnuts covered by a woody cap	Eucalypt
Species	*camaldulensis*	• Can grow up to 45 m tall • Pointed gum nut	River red gum

*Plant classification uses division instead of phylum.

▲ FIGURE 6.4.4 A river red gum is classified as *Eucalyptus camaldulensis*.

ACTIVITY

Modelling Linnaean classification

Models and analogies are useful to explain complex problems. In this task you are going to use an analogy to model the Linnaean classification of living things.

You need
- pencil case with a wide range of objects
- pictures of items commonly found in a kitchen (e.g. large plate, small plate, mug, glass, bowl, knife, spoon, fork, teaspoon, measuring cup, measuring spoon, oven, microwave, stove, mixing bowl, saucepan, wooden spoon, cake tin)
- pictures of items in a supermarket (e.g. apple, orange, banana, orange juice, milk, soft drink, bread, soup, rice, pasta, washing powder, dishwashing liquid, cleaning spray, shampoo, conditioner, hand soap, tissues, notebook, cat food, dog food, chocolate, potato chips)
- device to record a video (if possible)

What to do

Work in groups of two or three.

1. Choose one analogy for Linnaean classification.
    - Items in a pencil case
    - Items in a kitchen
    - Items in a supermarket
2. Discuss how these items can be classified into a hierarchy similar to the Linnaean classification.
3. Discuss how your analogy demonstrates how the items in a group become more similar as you move down the hierarchy, but there are fewer items in each group.
4. Model this classification, using your items or photos.
5. Plan and create a method of sharing your analogy and model with other members of the class. For example, create a video or a poster, or do a short oral presentation.

What do you think?

In what ways did your analogy and model accurately reflect the Linnaean classification of living things and in what ways was it inaccurate?

6.4 LEARNING CHECK

1. **List** the classification levels, in order.
2. Which level of classification contains groups with the most:
    a. similarities?
    b. organisms?
3. Roses and plum trees both belong to the Rosaceae family but they belong to different genera. Would they be more similar or less similar than two plants in the same genera? **Explain** your answer.
4. Conduct research into the taxonomy of the bottlenose dolphin. **State** the name of the group it is classified in for each level of classification.

6.5 The kingdoms of living things

BY THE END OF THIS MODULE, YOU WILL BE ABLE TO:
- ✓ list the five kingdoms of living things
- ✓ describe the key characteristics of each kingdom.

GET THINKING

In Linnaeus's original classification, there were only two kingdoms: plants and animals.
1. Why do you think Linnaeus only included these two kingdoms originally?
2. Why do you think there are now more kingdoms?

Classifying organisms into kingdoms

Organisms are classified into kingdoms based on key characteristics that are constant over the organism's life. These include:

- the number of cells in the organism (**unicellular** or **multicellular**)
- whether the cells are simple or complex
- how the organism obtains its food.

unicellular
composed of only one cell

multicellular
composed of many cells

Kingdom Animalia

You can easily identify something as an animal, including birds, fish, humans, snakes, worms, spiders and jellyfish. But what characteristics are common for all animals?

For an organism to be classified as an animal, it must be made of more than one cell (multicellular) and the cells must be complex. This means that they contain structures that carry out certain functions.

Animals are known as **heterotrophs** because they cannot make their own food. Instead, they must ingest (eat) their food to gain the energy and chemicals needed.

heterotroph
an organism that cannot make its own food and so gains nutrition by ingesting other sources

▲ **FIGURE 6.5.1** A rhinocerous is classified in kingdom Animalia.

Quiz
Kingdoms

Interactive resource
Match: Kingdoms

Kingdom Plantae

Plants include grasses, trees, shrubs, ferns and mosses. They have many structural similarities to animals because they are made up of many complex cells. However, plants are **autotrophs** because they can produce their own food through a process called **photosynthesis**. This is possible because plant cells have specialised structures that can use the Sun's energy to produce sugar.

autotroph
an organism that can make its own food

photosynthesis
the process by which plants use light energy from the Sun to produce simple sugars (e.g. glucose) in a series of chemical reactions

▲ **FIGURE 6.5.2** Plants use the Sun's energy to make their food.

Kingdom Fungi

The fungi kingdom includes organisms such as mushrooms, toadstools, yeasts and moulds. They are made of complex cells enclosed by a strong cell wall containing a substance called chitin. Most fungi are multicellular; however, yeasts are unicellular (Figure 6.5.3).

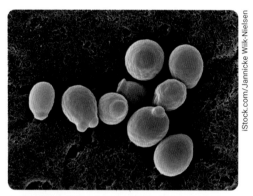

▲ **FIGURE 6.5.3** A magnified photo of yeast taken with a scanning electron microscope. Each yeast is made of only one cell.

▲ **FIGURE 6.5.4** Mushrooms are multicellular fungi.

Although fungi have many similar characteristics to plants, their way of obtaining food is different. Fungi cannot make their own food. Instead, they get their nutrition from decaying matter; this makes them heterotrophs. Fungi are different from other heterotrophs, including animals, because they use chemicals to break their food down before absorbing it.

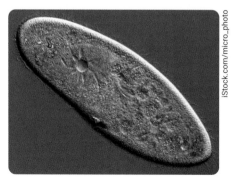

▲ **FIGURE 6.5.5** *Paramecium caudatum* is a species in the kingdom Protista.

Kingdom Protista

The kingdom Protista contains all organisms made of complex cells that are not fungi, plants or animals. Because of this broad definition, there is a lot of variation within this kingdom. However, most protists are unicellular and microscopic.

Some protists, such as green algae, are autotrophs. Others, such as paramecium, are heterotrophs (Figure 6.5.5).

Kingdom Bacteria (also known as kingdom Monera)

There are more than 30 000 named species of bacteria, including *Escherichia coli* (*E. coli*), *Streptococcus pyogenes* (Strep) and *Staphylococci epidermidis* (Staph). Bacteria are found everywhere: in soil, water, plants and animals, and within the earth. Some bacteria are harmful, but most are beneficial because they support organisms. For example, there are up to 1 000 000 000 000 000 (10^{15}) bacteria in the digestive system of a horse. The bacteria break down food so that the horse can absorb the nutrients.

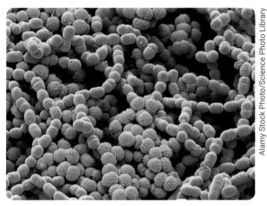

▲ **FIGURE 6.5.6** Bacteria seen through a scanning electron microscope

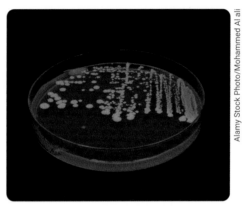

▲ **FIGURE 6.5.7** Bacteria can be grown on agar plates. Each dot on the plate contains millions of bacteria cells.

All bacteria are unicellular organisms with a simple cell structure. They are so small that between 20 000 and 1 million bacteria lined up in a row would be just 1 metre in length!

With such a large number of species of bacteria, it is easy to understand that there is great diversity between them. This includes their source of nutrition, as some are heterotrophs while others are autotrophs.

Other kingdoms

As scientists gain new knowledge and understanding, they adapt their theories and models. For example, the development of the scanning electron microscope allowed us to see things in much greater detail than ever before. Based on this new information, some scientists have proposed dividing living things into more kingdoms.

One suggestion is a sixth kingdom, kingdom Archaebacteria (or kingdom Archaea) containing simple, unicellular organisms that are different from bacteria. These organisms can survive in extreme environments such as those with very high salinity, very high temperatures and high acidity. When this sixth kingdom is included, the original kingdom Bacteria (or Monera) is divided into kingdom Eubacteria and kingdom Archaebacteria. Archaebacteria are considered to be primitive bacteria and Eubacteria are considered more recent.

Summarising the characteristics of the six kingdoms

Table 6.5.1 summarises the characteristics of the organisms in the six kingdoms.

TABLE 6.5.1 A summary of the characteristics of organisms in the six kingdoms

Kingdom	Number of cells	Complexity of cells	Method of nutrition
Animalia	Multicellular	Complex	Heterotroph
Plantae	Multicellular	Complex	Autotroph (photosynthesis)
Fungi	Unicellular or multicellular	Complex, contain chitin in the cell wall	Heterotroph (decaying matter)
Protista	Unicellular	Complex	Autotroph or heterotroph
Eubacteria	Unicellular	Simple	Autotroph or heterotroph
Archaebacteria	Unicellular	Simple, survive in extreme conditions	Autotroph or heterotroph

6.5 LEARNING CHECK

1. **Name** the kingdoms that include organisms with a complex cell structure.
2. **List** the characteristics of organisms in the plant kingdom.
3. **Name** the kingdom with organisms that are made up of many, complex cells and who cannot make their own food.
4. **Explain** why an organism would be classified as an animal and not a plant.
5. *Euglena* are single-celled organisms found in many aquatic environments. Their cells are quite complex, including the presence of a whip-like extension called a flagellum. *Euglena* also contain structures that allow them to undergo photosynthesis. However, they can also absorb food from the environment. What kingdom would *Euglena* be classified in? **Justify** your answer.
6. Originally, fungi were classified in the plant kingdom.
    a. **Suggest** why they were originally thought to be plants.
    b. **Explain** why they can no longer be classified as plants.
7. Conduct research to identify the kingdom for each of the following organisms.
    a. Diatom
    b. Dugite
    c. Anthrax
    d. Gardenia
    e. *Penicillium*
8. Each kingdom is divided into phyla or divisions. Use the Internet to **investigate** the phyla in the animal kingdom. What is the name of each phylum and what type of animal does it contain?

6.6 Naming living things

BY THE END OF THIS MODULE, YOU WILL BE ABLE TO:
- ✓ define binomial nomenclature
- ✓ describe how organisms are named.

GET THINKING

Table 6.6.1 shows the common and scientific names for some well-known organisms. What do you notice about all the scientific names and how they are written? List three things that they have in common. As you work through this module, reflect on your answer – did you predict the rules for writing scientific names?

▼ TABLE 6.6.1 The common and scientific names of some well-known organisms

Common name	Scientific name
Human	*Homo sapiens*
Domestic cat	*Felis catus*
Venus flytrap	*Dionaea muscipula*
Button mushroom	*Agaricus bisporus*

Scientific names

The word for 'cat' is different in different languages: *katze* in German, *chat* in French, *gatto* in Italian, *katinas* in Lithuanian and *kath* in Welsh. It makes it difficult to talk about cats if you are not speaking the same language. However, if you ask an English, German, French, Italian, Lithuanian or Welsh scientist the scientific name for a cat, the answer will be the same: *Felis catus*.

Video activity
Naming species

Extra science investigations
Identifying insects using a dichotomous key
Lolly dichotomous key

▲ FIGURE 6.6.1 The scientific name for a domestic cat is *Felis catus*.

Binomial nomenclature

In Table 6.6.1, you would have noticed that the genus name for humans is *Homo*, and the name at the species level (the **specific name**) is *sapiens*. The scientific name for humans is made up of the genus name together with the specific name – *Homo sapiens*. This is also shown in Table 6.4.1, where the genus name for the river red gum is *Eucalyptus* and the specific name is *camaldulensis*; hence, the scientific name for the river red gum is *Eucalyptus camaldulensis*. This two-word naming system is known as **binomial nomenclature**.

specific name
the second part of the scientific name that identifies the species within a genus

binomial nomenclature
a two-word naming system for naming living things

Chapter 6 | Classifying living things

Naming conventions

When writing scientific names, there are conventions, or rules, to follow.

1. The scientific name is in italics. This is because it is in Latin and also distinguishes it from other text. If you are handwriting the names, underline them instead.
2. The genus name is always written first and begins with a capital letter. The species name is always written second and begins with a lower-case letter.
3. If the same scientific name occurs more than once in the same piece of writing, the genus name may be abbreviated to its first letter, such as *H. sapiens* or *E. camaldulensis*.

Comparing scientific names

Within a genus, the species are often distinguished with descriptive names. For example, there are approximately 900 different species in the *Eucalyptus* genus. The scientific name for each species begins with *Eucalyptus*, but each one has a different specific name to identify it. *Eucalyptus longifolia* is named for its long (*longus*) leaves (*folium*), and *Eucalyptus grandis* is named for its large size (*grandis*) (Figure 6.6.2).

▲ **FIGURE 6.6.3** **(a)** *Eucalyptus longifolia* is named for its long leaves, whereas **(b)** *Eucalyptus grandis* is named for its large size.

This method means that the same specific name can be used for species in more than one genus. However, each species will have its own, unique scientific name because it is the combination of the genus and specific name that identifies each species. For example, the New Zealand iris is named *Libertia grandiflora* because *grandiflora* means large flowers. The Bull Bay magnolia also has large flowers, and its scientific name is *Magnolia grandiflora*. Even though both plants have the same specific name, they are in different genera. The lily magnolia (*Magnolia liliiflora*) is in the same genus as the Bull Bay magnolia. This means that the Bull Bay magnolia will have more in common with the lily magnolia than the New Zealand iris, which is in a different genus.

Naming new species

When a new species is discovered, its kingdom, phylum and so on are determined on the basis of characteristics of groups that are already determined. In some cases, it will belong in a genus that already exists, and the new species simply needs a new specific name. For example, in 2021, scientists in Australia identified a new species of tree frog belonging to the genus *Litoria*. The species, *Litoria quiritatus*, is named for its high-pitched call, as *quirito* means a shriek or scream in Latin.

Weblink
The screaming tree frog

▲ **FIGURE 6.6.3** The screaming tree frog (*Litoria quiritatus*) was identified and named in 2021.

Other organisms will not fit into predetermined groups, and so scientists make a new group. This was the case with a new species of brittle star discovered in 2015 by Australian scientist Tim O'Hara. O'Hara found the organism in specimens that had been collected in the south-west Pacific Ocean near New Caledonia. The new brittle star has eight snake-like arms with rows of hooks and spines, and a row of teeth along every jaw. It is so different from other brittle stars that a new family and genus was created for it. It was named *Ophiojura exbodi* because it was collected during the EXBODI expedition.

▲ **FIGURE 6.6.4** *Ophiojura exbodi* is a new species of brittle star.

ACTIVITY

Research activity

Scientific names of organisms found in Australia

Figure 6.6.5 shows the common names and photos of some organisms found in Australia. Your task is to research the scientific name for each organism, where they are found and their key features. Use this information to write a paragraph with a possible explanation for why each one was given that particular name.

▲ **FIGURE 6.6.5** Some Australian organisms

6.6 LEARNING CHECK

1. How many words are in the scientific name of an organism?
2. Why are scientific names written in italics?
3. **Explain** why the scientific name for an organism is the same around the world.
4. Tapeworms have a ribbon-like body composed of a series of segments. The different species of tapeworm vary in the organism they infect.

   a **Apply** the information in Table 6.6.2 to write the scientific name for this tapeworm.

   ▼ **TABLE 6.6.2** The Linnaean classification for a tapeworm

Level of classification	Name
Kingdom	Animalia
Phylum	Platyhelminthes
Class	Cestoda
Order	Cyclophyllidea
Family	Taeniidae
Genus	*Taenia*
Species	*ovis*

   b What name indicates which organism this tapeworm infects?

   c **Name** the organism that this tapeworm infects. You may need to do some research to find this answer.

6.7 First Nations Australians' classification systems

FIRST NATIONS SCIENCE CONTEXTS

IN THIS MODULE YOU WILL:
✓ examine First Nations Australians' classification systems.

Classification systems

For thousands of years, First Nations Australians have observed and learned from the complex interrelationships between organisms in their natural environments, including relationships with other First Nations Peoples. They have used these observations to develop culturally specific and complex classification systems.

The different classification systems developed by many First Nations Peoples across Australia reflects the diverse environments each Nation encompasses. However, there are many similarities in the systems that First Nations Australians use to classify living things. For example, many classification systems include criteria relating to an organism's physical characteristics, behaviour and habitat. Sub-classifications may incorporate life cycle stage, sex, age or reference to a particular custom or practice.

Many First Nations Australians classify living organisms into plant or animal groups. Animal life may then be divided into groups based on observable characteristics and habitats, such as land animals, marine animals and winged organisms. Similar to the Linnaean system of classification, many First Nations Australians' classification systems are hierarchical, with organisms grouped in levels.

Classifying animals

The Yolngu Peoples of north-eastern Arnhem Land in the Northern Territory have a highly structured system for classifying plants and animals in which animals showing similar features are grouped together. For example, birds are placed in a group called *warrakan*. Within this group is a subgroup for small birds, called *djikay*. Within this subgroup is another smaller group for finches, called *lidjilidji* (see Figure 6.7.1). In this system of classification, each large group is made up of all the subgroups below it.

▲ FIGURE 6.7.1 The Gouldian (rainbow) finch is part of the *lidjilidji* group within the Yolngu classification system. It was found in many northern areas across Australia but is now identified as endangered.

▲ FIGURE 6.7.2 A dugong mother and calf

The Yanyuwa Peoples of the Northern Territory classify living organisms as being either coastal and marine or inland. Organisms may be further sub-classified based on the particular habitat they occupy, such as open sea or intertidal zone. The Yanyuwa Peoples' language has one term, *walya*, that refers to all dugong and sea turtles. This category further subdivides to include 16 different names to distinguish between dugongs based on age, size, gender and status within its herd. These classifications indicate the importance of this animal to the Yanyuwa Peoples.

Classifying plants

The importance of plants as a resource for First Nations Australians is reflected in their classification systems. Many classification systems distinguish plants as either wood-bearing or non-woody, highlighting the significance of wood in First Nations Australians material culture. For example, the Anindilyakwa Peoples of Groote Eylandt distinguish between woody plants (*eka*) and other plants (*amarda*) in the region. The woody plants are then sub-classified into a further eight categories based on the observation of a plant's form or habitat.

The classification of wood-bearing plants may also be based on function and use. In this classification, the wood-bearing plant may have a name based on the function of a finished object such as spear trees, shield trees, canoe trees or resin trees, and many other use-based categories. For example, the Pitta Pitta Peoples of the Boulia region in Queensland classify both the tree *Erythrina vespertilio* and the shields constructed from it as *koon-pa-ra*.

▲ FIGURE 6.7.3 This tree on Kaurna Country bears a scar from where a shield has been harvested from it.

Sharing knowledge

First Nations Australians' detailed plant and animal knowledge and classification systems are instrumental in providing knowledge to scientists as they apply the Linnaean classification systems to the plants and animals of Australia. There are many instances where scientists thought they had discovered a 'new' species, only to be told the already existing names, cultural stories and detailed understanding of the 'new' species by First Nations Australians.

ACTIVITY

1. **Describe** the similarities between First Nations Australians' and Linnaean classification systems.
2. **Construct** a dichotomous key to represent the Yolngu People's classification system of birds.
3. **Explain** the advantages of classifying living things based on function and use. Provide an example of where you could classify local plants on this basis.

6.8 Changing classifications

SCIENCE AS A HUMAN ENDEAVOUR

BY THE END OF THIS MODULE, YOU WILL BE ABLE TO:
✓ explain how and why the classification of living things changes over time.

Changes at the kingdom level

Science is all about making observations, asking questions, developing an understanding of our world and using this understanding to solve problems. In any area of science, as people gain new information, scientists reflect on existing theories and models. The new information is incorporated to affirm or adapt what we previously believed.

Video activity
What is taxonomy?

This cycle occurs in taxonomy, leading to changes in classification. In Linnaeus's time, the first grouping was whether a living thing was a plant or an animal. If it was a plant, it was classified in the kingdom Plantae; if it was an animal, it was classified in the kingdom Animalia. With the search for more knowledge and the invention of microscopes, which enabled scientists to see cells, it was not so easy to classify living things into one of these two groupings. In 1866, a third kingdom, Protista, was proposed. This kingdom contained organisms made up of only one cell. As you learned in this chapter, it is now commonly accepted that there are six kingdoms. However, some scientists believe that there is evidence for up to eight kingdoms!

Changes to ape classification

Changes in classification also occur at levels lower in the hierarchy. We now have technology that allows us to look at an organism's genetic code. This depth of information provides greater detail than was previously available about the relationship between organisms.

One example of this change is in the apes, including humans. Originally, apes were grouped into three families:
- humans
- chimpanzees, gorillas and orangutans
- tree-dwelling apes such as the gibbon.

We now know that humans and chimpanzees are more closely related than this classification represented and so the classification was adjusted. Figure 6.8.1 represents the new classification of apes. Note that it includes the subfamily and tribe groups. These are further divisions of the Linnaean classification levels.

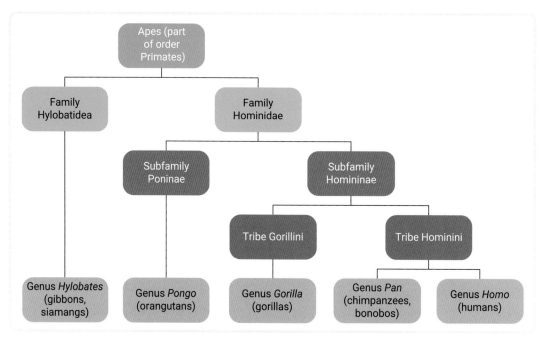

▲ FIGURE 6.8.1　The updated classification of apes

▲ FIGURE 6.8.2　Humans and chimpanzees are closely related, which is reflected in the updated classification of these species.

6.8 LEARNING CHECK

1. **Describe** how the number of kingdoms of living things has changed since Linnaeus first proposed his classification system.
2. **Describe** why classifications may change.
3. **Explain** why the number of ape families was changed.
4. Do you think there will be more, or fewer, changes in classification now than 50 years ago? **Discuss** the reason for your answer.

SCIENCE INVESTIGATIONS

6.9 Creating a dichotomous key

SCIENCE SKILLS IN FOCUS

IN THIS MODULE, YOU WILL FOCUS ON LEARNING AND IMPROVING THESE SKILLS:

- create a dichotomous key by grouping items based on their characteristics.

In this chapter, you have learned about dichotomous keys. In this activity, you are going to create a dichotomous key for science equipment.

Here are some pointers to remember when creating a dichotomous key.

1. Dichotomous keys are used to identify objects or living things.
2. Before you start making your key, you need to know the properties of the objects that you want to classify. The properties that you use in the dichotomous key should be objective traits that do not change. This means that the object is consistently classified in the same way.
3. At each step in the key, you need to have two options. No more, no less!
4. Each option needs to be written as a statement (a fact), not a question.

For example:
3 cm or bigger
Smaller than 3 cm

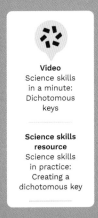

Video
Science skills in a minute: Dichotomous keys

Science skills resource
Science skills in practice: Creating a dichotomous key

SCIENCE EQUIPMENT DICHOTOMOUS KEY

AIM

To create a dichotomous key (tree diagram and linked versions) for science equipment

YOU NEED

- ☑ pictures of each of the pieces of science equipment shown in Figure 6.9.2. Alternatively, your teacher may make the actual equipment available to you
- ☑ large piece of paper or a whiteboard
- ☑ pens

WHAT TO DO

Work in groups of two or three.

PART A: CREATING A TREE DIAGRAM FOR A DICHOTOMOUS KEY

1. List the characteristics of each piece of equipment.
2. Organise the equipment into two groups based on similar characteristics. Remember to use characteristics that don't change and are not based on subjective opinion.
3. Represent this organisation at the start of a tree diagram. An example is shown in Figure 6.9.1.

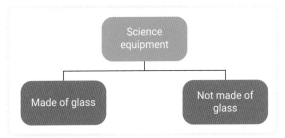

▲ FIGURE 6.9.1 The start of a dichotomous key for science equipment

4. Choose one of your groups. Divide it into two smaller groups based on similar characteristics.
5. Add this organisation to your tree diagram. An example is shown in Figure 6.9.3.
6. Repeat steps 4 and 5 until you have only one item in the group. Then, write the name of the item below the characteristic. An example is shown in Figure 6.9.4.

Chapter 6 | Classifying living things

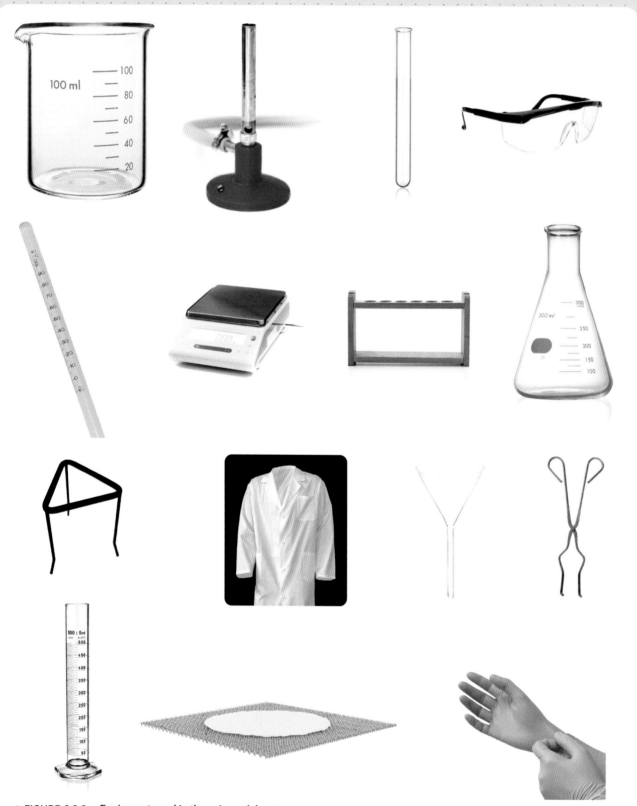

▲ FIGURE 6.9.2   Equipment used in the science laboratory

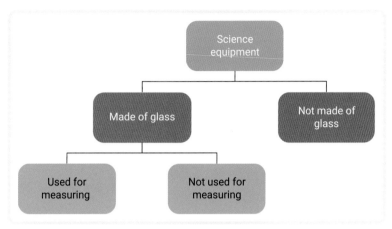

▲ **FIGURE 6.9.3** The continuation of a dichotomous key for science equipment

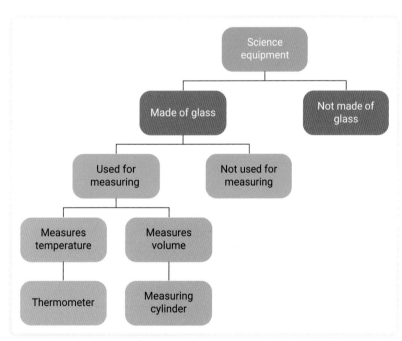

▲ **FIGURE 6.9.4** A further continuation of a dichotomous key for science equipment

7 Repeat step 6 for each group until there is only one item in every group.

PART B: CREATING A LINKED KEY FOR A DICHOTOMOUS KEY

1 Using your tree diagram, write a number at each division of a group. These will become the numbers for your descriptions.

2 Start with number 1. Write the description for option a, and the description for option b underneath each other.

3 After the description, write what number that option should go to. This is written at the right-hand side of the line. Often a series of dots are used to connect the description and direction. Figure 6.9.5 shows an example.

4 Continue this process until you have written the description for each option.

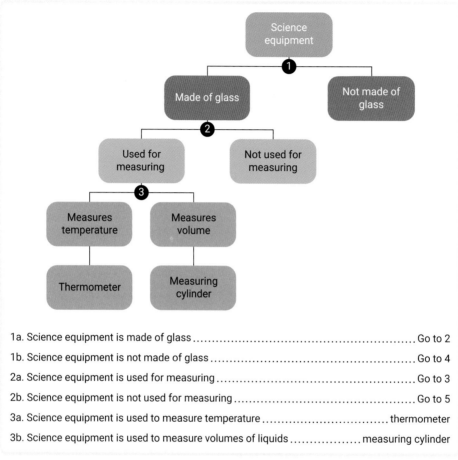

▲ FIGURE 6.9.5 Starting a linked key (tabular key) for a dichotomous key for science equipment

RESULTS

If possible, take a photo of your group's dichotomous keys.

WHAT DO YOU THINK?

1. Were some items hard to classify? What made them hard to classify?
2. Do you think you could have done this activity last year? Explain your answer.
3. Were the dichotomous keys from other groups the same as yours? Explain your answer.
4. If you were to do this activity again, what would you change to make the process more effective?
5. What challenges do you think scientists would face when creating dichotomous keys for living organisms?
6. Use this process to create a dichotomous key for a range of lollies.

CONCLUSION

Summarise what a dichotomous key is and what you did to create one for science equipment.

6 REVIEW

REMEMBERING

1. What are we doing when we classify something?
2. What does the 'di' in dichotomous mean?
3. What are the classifications in the Linnaean system based on?
4. **List** the seven levels of classification in order of increasing similarity.
5. **Define** 'species'.
6. **List** three features shared by all members of the kingdom Animalia.
7. **Name** the smaller groups that a kingdom is divided into.

UNDERSTANDING

8. **State** two reasons why we classify objects.
9. What are the best types of features on which to base a classification system? **Justify** your answer.
10. How does the Linnaean classification system benefit scientists around the world?
11. Are organisms in the same genus more or less related than organisms in the same:
    a. class?
    b. species?
12. **Explain** where the scientific name for a species comes from.
13. Why are structural features used to classify organisms, rather than their colour?

APPLYING

14. **Apply** your knowledge of classification to create a classification system for these objects: wooden ruler, eraser, biro, black lead pencil, pencil sharpener, felt pen, protractor.
15. **Describe** a situation in which classification affects your life.

16. The first edition of *Systema Naturae* had only 13 pages. The 13th edition was published in 1770 and had 3000 pages. Why do you think the book increased in size so much?
17. Could the members of the kangaroo species *Macropus greyii* and *Macropus rufus* interbreed and produce fertile offspring? **Explain** your answer.
18. Use a search engine to find the scientific names for the following organisms. State the scientific name and the abbreviated form of that name and **explain** how the name was chosen for that species.
    a. Sturt desert pea
    b. Firewood banksia
    c. California sea hare
    d. Spectacled bear
19. **Explain** why it would be important for a farmer to breed using only one species of wheat.
20. Why are *Euglena* and slime moulds placed into the kingdom Protista?
21. **Explain** how the species name *Ornithorhynchus anatinus* complies with the rules of binomial nomenclature.
22. Are members of the species *Aspergillus niger* more closely related to *Helleborus niger* or to *Aspergillus clavatus*?

    Explain the reasons for your answer.
23. Give an example of one First Nations Australians' classification system. **Describe** the basis on which the organisms are classified.

EVALUATING

24. Find two places in your home where a classification system is used. **Explain** what the system is, and how it assists in organising the items. Can you think of a better system to use?
25. **Explain** how our own names (first name and surname) could fit the rules of binomial nomenclature. How is our naming system similar to binomial nomenclature? How is it different?

26 The following table shows the common and scientific names of some cats.

▼ The big cats

Common name	Scientific name
Bobcat	Lynx rufus
Lynx	Lynx canadensis
Mountain lion	Puma concolor
African lion	Panthera leo
European wildcat	Felis sylvestris
Tiger	Panthera tigris
Jaguar	Panthera onca

a How many genera are represented in the table?
b How many different species are represented in the table?
c The scientific name of the domestic cat is *Felis catus*. Which cats in the table are most closely related to the domestic cat?

27 **Suggest** why the Linnaean system of classification doesn't include breeds.

28 Could an organism's diet be used as a basis for a classification system? **Explain** why or why not.

29 Classification systems also indicate how closely related objects within the system are. **Explain** if, and how, the Linnaean system achieves this.

CREATING

30 **Create** an imaginary organism from any kingdom. Draw a labelled diagram to represent your organism. Swap imaginary organisms with a partner. Classify each other's organisms according to the Linnaean classification system.

31 **Create** a tree diagram and a tabular form of a dichotomous key for different breeds of dogs.

BIG SCIENCE CHALLENGE PROJECT

1 Connect what you've learned

Observe the photos of a range of organisms. Think about what you have learned in this chapter that relates to the photos.

- What kingdoms are they classified in?
- How would they be named?
- What properties would be used to classify them?
- How could a dichotomous key be used to classify them?

2 Check your thinking

At the start of the chapter, you answered the following questions about the organisms in the photos.

a What do they all have in common?
b What are the main characteristics of each?
c How can they be organised into groups?
d How are they named?

Use the understanding that you have gained in this chapter to improve your answers. This process is similar to how scientists refine their understandings as they gain new knowledge.

3 Make an action plan

Use the characteristics of the organisms to develop a dichotomous key that could be used by another student to identify the name of each organism.

4 Communicate

Present your dichotomous key as one of the following.

- Tree diagram on a poster
- Linked (tabular) key in an A4 pamphlet
- A PowerPoint or Keynote presentation that uses links to move to the next choice

7 Ecosystems

7.1 What is an ecosystem? (p. 188)
An ecosystem includes the living and non-living factors in an environment, and how they interact.

7.2 Energy in an ecosystem (p. 190)
Photosynthesis and cellular respiration allow energy to enter or leave an ecosystem.

7.3 Reviewing food chains (p. 192)
Food chains model the energy flow from one organism to another.

7.4 Food webs (p. 195)
Food webs represent energy flow through an ecosystem.

7.5 Movement of energy and matter in an ecosystem (p. 200)
Matter cycles through an ecosystem.

7.6 Modelling ecosystems with pyramids (p. 202)
Pyramids can visually represent the biomass, energy or number of organisms in an ecosystem.

7.7 FIRST NATIONS SCIENCE CONTEXTS: First Nations Australians' traditional ecological knowledge (p. 206)
First Nations Australians have a deep understanding of the ecosystems in which they live.

7.8 SCIENCE AS A HUMAN ENDEAVOUR: Human impact – introduced species (p. 209)
Changes to an ecosystem, such as introducing a new species, can have a significant impact.

7.9 SCIENCE INVESTIGATIONS: Science communication (p. 212)
Understanding an environmental issue

BIG SCIENCE CHALLENGE #7

▲ **FIGURE 7.0.1** The Wooroloo bushfire had a devastating impact on plants, animals and ecosystems.

The Wooroloo bushfire on 1 February 2021 destroyed 10 900 hectares of bushland in the Perth Hills. Think about the factors that would affect the plants and animals:

▶ during the fire

▶ immediately after the fire

▶ a year after the fire.

Summarise your thoughts in a suitable format, such as a table, mind map or annotated diagram.

#7 SCIENCE CHALLENGE ACCEPTED!

At the end of this chapter, you can complete Big Science Challenge Project #7. You can use the information you learn in this chapter to complete the project.

Assessments
- Prior knowledge quiz
- Chapter review questions
- End-of-chapter test
- Portfolio assessment task: Research project

Videos
- Science skills in a minute: Science communication **(7.9)**
- Video activities: Biotic factors in ecosystems **(7.1)**; Abiotic factors in ecosystems **(7.1)**; What is an ecosystem? **(7.1)**; Photosynthesis **(7.2)**; Oceanic food chains **(7.3)**; Cane toads **(7.8)**

Science skills resources
- Science skills in practice: Science communication **(7.9)**
- Extra science investigations: Light and photosynthesis **(7.2)**; Owl pellets **(7.3)**; Who is in the pond? **(7.4)**

Interactive resources
- Label: Photosynthesis v cellular respiration **(7.2)**; Biomass pyramid **(7.6)**
- Drag and drop: Build a food web **(7.4)**
- Match: Food chains **(7.3)**

Nelson MindTap

To access these resources and many more, visit **cengage.com.au/nelsonmindtap**

Chapter 7 | Ecosystems

7.1 What is an ecosystem?

BY THE END OF THIS MODULE, YOU WILL BE ABLE TO:
- ✓ define ecosystem, biotic factor, abiotic factor, community and population
- ✓ describe and explain the interactions in an ecosystem.

> **GET THINKING**
>
> Think about a bird in the garden. Make a list of all the things that may have affected the bird in the last day. Write these in two columns – the things that are living and the things that are non-living. These things, or factors, and how they interact with one another are the ecosystem in which the bird lives.

environment
a unique set of non-living and living factors for a particular area and time

biotic factor
a living component of an ecosystem

population
organisms of one species living together

community
all the organisms that live together and interact

abiotic factor
a non-living component of an ecosystem

Environments

An **environment** is the set of conditions within a given area. This includes the:
- living components (**biotic factors**) such as all the plants, animals, bacteria and fungi. These biotic factors include **populations** of different organisms, which combine to make up the **community** within the environment
- non-living components (**abiotic factors**) such as temperature, amount of light, rainfall, pH of soil and saltiness (salinity) of water.

Australia has many different types of environments. These range from the hot, humid environments in tropical Queensland to the cold, snowy environments of the Victorian alpine region and the hot, dry environments of Central Australia.

▲ **FIGURE 7.1.1** Australia has many different types of ecosystems: **(a)** Great Barrier Reef, **(b)** Kakadu National Park, **(c)** Simpson Desert, **(d)** Daintree Rainforest, **(e)** Murray River red gum forest, **(f)** alpine ecosystem in Tasmania.

Ecosystems

ecosystem
the living and non-living factors of an environment and all their interactions

An **ecosystem** is defined as the biotic and abiotic factors, and their interactions, in a certain place. It is slightly different from an environment because it also includes how the factors interact with, or affect, one another.

The sheep in Figure 7.1.2 are part of a paddock ecosystem. Some of the environmental factors are listed in Table 7.1.1. These factors interact with one another in the ecosystem. For example, the Sun's energy is used by the grass, the sheep eat the grass, the trees provide shade and protection for the sheep, the worms get their nutrition from the sheep manure and the moisture in the soil is used by the trees and grass.

▼ TABLE 7.1.1 Factors that make up the environment in the paddock

Biotic factors	Abiotic factors
Sheep	Shade from trees
Grass	Light
Trees	Temperature
Flies	pH of soil
Birds	Water in soil
Bacteria in soil	Water in troughs or dams for the sheep
Bacteria in sheep	Humidity
Worms	Wind

▲ FIGURE 7.1.2 Sheep in a paddock ecosystem

7.1 LEARNING CHECK

1 **Define** 'ecosystem'.
2 **List** two biotic and two abiotic factors in your classroom environment.
3 **Explain** how an ecosystem is different from an environment.
4 Figure 7.1.3 shows a marine ecosystem.
   a **List** three biotic factors in this ecosystem.
   b **List** three abiotic factors in this ecosystem.
   c **Describe** two interactions that would occur in this ecosystem.
5 There are many different types of ecosystems. However, they can all be classified as terrestrial, freshwater or marine.
   a **Describe** the key characteristic(s) of each of these types of ecosystems.
   b Do you think all terrestrial ecosystems are the same? **Discuss** your answer, including evidence for your reasoning.
6 Find an area outside, away from buildings, cars and people. Sit quietly and observe the ecosystem.
   a **List** as many biotic and abiotic factors as you can.
   b **Describe** five interactions between these factors.
   c **Model** the ecosystem in a diagram. Use labels and arrows to indicate the factors and interactions.
   d How do you think the environment is different today from 100 years ago? What do you think caused the change? **Discuss** your answers.

▲ FIGURE 7.1.3 A marine ecosystem

Video activities
Biotic factors in ecosystems
Abiotic factors in ecosystems
What is an ecosystem?

7.2 Energy in an ecosystem

BY THE END OF THIS MODULE, YOU WILL BE ABLE TO:
- ✓ model photosynthesis and cellular respiration with word equations
- ✓ explain the role of photosynthesis and cellular respiration in an ecosystem.

Video activity
Photosynthesis

Interactive resource
Label: Photosynthesis v cellular respiration

Extra science investigation
Light and photosynthesis

GET THINKING

Skim read this module, paying particular attention to the word equations for photosynthesis and cellular respiration. How do you think these processes allow energy to enter and move through an ecosystem?

Photosynthesis

The energy source for all ecosystems is the Sun. Plants contain a green pigment called **chlorophyll** that can absorb the Sun's energy and use it in a series of chemical reactions that produce sugars such as glucose. This process is called **photosynthesis**, which is summarised in a word equation in Figure 7.2.1.

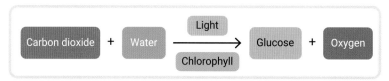

▲ **FIGURE 7.2.1** The word equation for photosynthesis

chlorophyll
a green pigment in plants that absorbs the Sun's energy; assists photosynthesis

photosynthesis
the process by which plants use light energy from the Sun to produce simple sugars (e.g. glucose) in a series of chemical reactions

During photosynthesis, carbon dioxide and water are converted to oxygen and simple sugars such as glucose. This allows plants to convert the light energy from the Sun into chemical energy stored in the simple sugars. Because plants can produce their own sugar in this way, they are called autotrophs. When other organisms eat plants, they consume this energy-rich sugar. Organisms that get their energy from consuming other organisms are known as heterotrophs.

▲ **FIGURE 7.2.2** Plants convert the Sun's energy into chemical energy during photosynthesis.

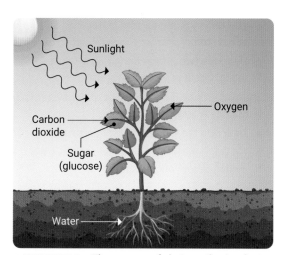

▲ **FIGURE 7.2.3** The process of photosynthesis: plants take in carbon dioxide from the air and water and use light energy to produce sugars and oxygen.

Cellular respiration

The energy stored in sugars can be released during a process called **cellular respiration**. As the name implies, cellular respiration occurs in all cells, including those of plants and animals. During cellular respiration, sugar (e.g. glucose) and oxygen are converted to carbon dioxide and water, releasing energy. This energy can be used for any active process, including growth, movement, digestion, healing and thinking.

cellular respiration the process by which cells use simple sugars (e.g. glucose) to produce energy that the organism can use

▲ FIGURE 7.2.4 The word equation for cellular respiration

Plants use some of the sugars they produce during photosynthesis in cellular respiration. The rest remains in the plant as a store of energy. When an animal eats the plant, the energy stored in the sugars is available to the animal. That animal will use some of the energy and store the rest. The stored energy will be available to the animal that eats it.

▶ FIGURE 7.2.5 The horse and rider use energy released from cellular respiration.

7.2 LEARNING CHECK

1 **Write** the word equation for:
   a photosynthesis.
   b cellular respiration.

2 Copy and **complete** the table below by placing a tick (yes) or cross (no) to indicate whether the organism uses each process.

Process	Plants	Animals
Cellular respiration		
Photosynthesis		

3 **Compare** photosynthesis and cellular respiration by listing two similarities and two differences.

4 **Suggest** why we breathe faster when we exercise.

5 The roots of the word respiration are 're', meaning again, and 'spirare', meaning 'to breathe'. Therefore, respiration means 'to breathe again'. This term is applied to the act of breathing because we are taking oxygen in and carbon dioxide out of our body. **Discuss** how this also applies to the term 'cellular respiration'.

6 Do you think it would be possible for an ecosystem to survive without any light? **Discuss** the reason for your answer.

7.3 Reviewing food chains

BY THE END OF THIS MODULE, YOU WILL BE ABLE TO:
- ✓ apply appropriate terminology to describe organisms within a food chain
- ✓ explain the role of different organisms within a food chain
- ✓ model the flow of energy and matter in an ecosystem with food chains.

Video activity
Oceanic food chains

Interactive resource
Match: Food chains

Extra science investigation
Owl pellets

food chain
a single linear diagram that shows the way energy and matter are transferred from producer to consumers

producer
an organism that produces its own food; usually a plant; an autotroph

consumer
an organism that must consume its food; an animal; a heterotroph

primary consumer (first-order consumer)
an organism that eats a producer; herbivore or omnivore

herbivore
an organism that feeds solely on plants; a primary consumer

secondary consumer (second-order consumer)
an organism that eats a primary consumer; carnivore or omnivore

carnivore
an organism that feeds solely on animals; a meat-eater

tertiary consumer (third-order consumer)
an organism that eats a secondary consumer; carnivore or omnivore

apex predator
the organism at the top of a food chain

GET THINKING

In groups of two or three, brainstorm keywords that you remember about food chains from when you learned about them in primary school. Make flashcards, a Kahoot! or a Quizlet with the word and definition. As you work through this module, add new terms to your collections and refine your existing definitions.

Energy flow

In Module 7.2, you learned how:
- energy from the Sun enters an ecosystem via photosynthesis
- light energy is converted to chemical energy in sugars
- chemical energy passes from one organism to another when they eat the sugars.

This passing of energy from one organism to another allows energy to flow through an ecosystem. A **food chain** represents this flow of energy. Because the energy is contained in the food, a food chain also represents the flow of matter.

In a food chain, the organisms are listed in order, starting with the plant because this is where the energy enters the ecosystem and the food is produced. An arrow between the organisms shows the flow of energy and matter. The arrow points towards the next organism, indicating that the energy and matter are transferred into this organism. This is illustrated in a simple food chain in Figure 7.3.1.

▲ **FIGURE 7.3.1** A food chain

Roles in a food chain

Each organism plays an important role in a food chain.

A **producer** makes its own food, bringing energy into an ecosystem. This is why the producers are at the start of a food chain. In most cases, the producer is a plant. Algae, a protist, is an example of a producer that is not a plant.

Consumers eat their food to gain energy and, therefore, are animals. They may be a:
- **primary consumer** (or **first-order consumer**) that eats the producer. Primary consumers are **herbivores** because they eat plants
- **secondary consumer** (or **second-order consumer**) that eats the primary consumer. Secondary consumers are **carnivores** because they eat meat. They may also be omnivores
- **tertiary consumer** (or **third-order consumer**) that eats the secondary consumer, making it a carnivore or omnivore.

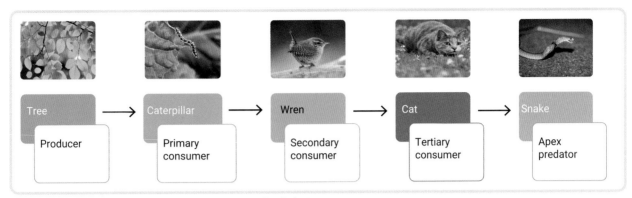

▲ FIGURE 7.3.2 Classification of the organisms in a food chain

The **apex predator** is the last consumer and the last organism in the food chain. Therefore, it is at the top of the food chain.

Some animals, called **omnivores**, eat plants and animals. Humans are omnivores because we can eat some plants, such as lettuce, but we can also eat meat, such as beef. Omnivores are primary consumers in food chains when they eat plants. They are higher-order (secondary or tertiary) consumers in food chains when they eat animals.

omnivore
an organism that eats both plants and animals

Trophic levels

Another way of classifying organisms in a food chain is by their **trophic levels**. A trophic level is the position in the food chain.

trophic level
the level or position in a food chain

- Producers make up the first trophic level.
- Primary consumers make up the second trophic level.
- Secondary consumers make up the third trophic level.
- Tertiary consumers make up the fourth trophic level.

ACTIVITY

Constructing a food chain

The following table gives names, photos and information about organisms in a food chain.

Photos and names				
Honey eater	Wedge-tailed eagle	Bee	Silver princess	Butcherbird
Descriptions				
A medium-sized black-and-white bird with a dark, hooked tip to its beak that it uses to kill its prey, such as small birds	A yellow-and-black insect that eats nectar and pollen	A large bird of prey that swoops down and carries its food away	A native Australian bird that eats nectar, insects and berries	A eucalyptus tree with large red flowers in autumn and winter

What to do

1 Match the description for each organism with the name and photo.
2 Use the descriptions to construct a food chain for the organisms.
3 Identify the:
   a producer.
   b first-order consumer.
   c apex predator.
   d secondary consumer.
   e second trophic level.

7.3 LEARNING CHECK

1 What types of food do each of the following eat?
   a Carnivore
   b Omnivore
   c Herbivore

2 **Match** each term with its description.

Term	Description
Secondary consumer	Makes its own food
Consumer	Eats the primary consumer
Producer	The last organism in a food chain
Apex predator	Must eat its food
Primary consumer	Eats the secondary consumer
Tertiary consumer	Eats a producer

3 The following food chain is for a farming ecosystem.
   Grass → sheep → dingo → eagle
   a **Name** the producer in the food chain.
   b What is the apex predator in the food chain?
   c How many consumers are in the food chain?
   d What organism is a herbivore in the food chain?
   e **Name** the secondary consumer in the food chain.

4 **Draw** a food chain in which:
   a algae are producers.
   b a shark is the apex predator.
   c there are three consumers.
   d an aphid is a primary consumer.
   e there is a cat in the food chain.

5 List one similarity and one difference between a:
   a primary and a secondary consumer.
   b producer and a consumer.
   c herbivore and an omnivore.

6 **Describe** what the arrows in a food chain represent.

7 Is it possible for a secondary consumer to be a herbivore? **Explain** your answer.

8 Are apex predators always in the fourth trophic level? Use examples to support your answer.

9 What organisms, other than plants, can fulfil the role of producer in a food chain? **Explain** your answer.

7.4 Food webs

BY THE END OF THIS MODULE, YOU WILL BE ABLE TO:
- ✓ define food web
- ✓ explain how a food web represents the flow of energy and matter through an ecosystem
- ✓ model the flow of energy and matter in an ecosystem with a food web.

GET THINKING

Discuss these questions with a partner.
1. Do you only eat one type of food?
2. Do you think organisms are part of only one food chain?
3. Why do you think your answers to Questions 1 and 2 are important in an ecosystem?

Interactive resource
Drag and drop: Build a food web

Extra science investigation
Who is in the pond?

Interconnecting food chains

An ecosystem contains a variety of organisms. Most organisms feed on more than one food source and often organisms are consumed by more than one animal. Therefore, there are many interconnecting food chains occurring in each ecosystem.

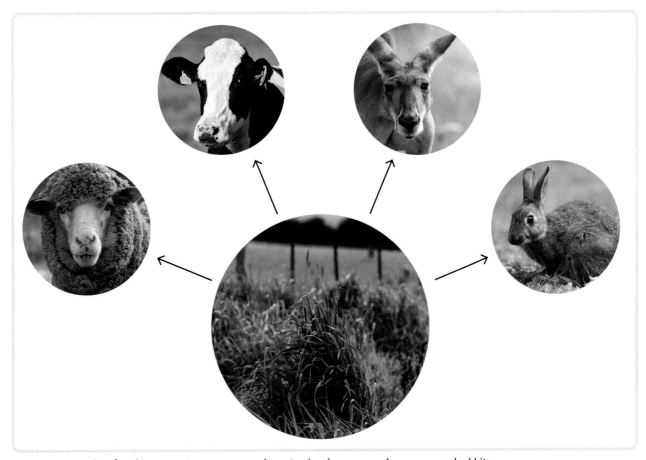

▲ **FIGURE 7.4.1** In a farming ecosystem, grass may be eaten by sheep, cows, kangaroos and rabbits.

Chapter 7 | Ecosystems

food web
a group of interlinked food chains that gives an overall picture of how energy and matter are transferred through an ecosystem

What is a food web?

A **food web** is all the food chains that occur in an ecosystem linked together. It gives a bigger picture of the flow of energy and matter through an ecosystem than a single food chain. Figure 7.4.2 shows an Australian food web.

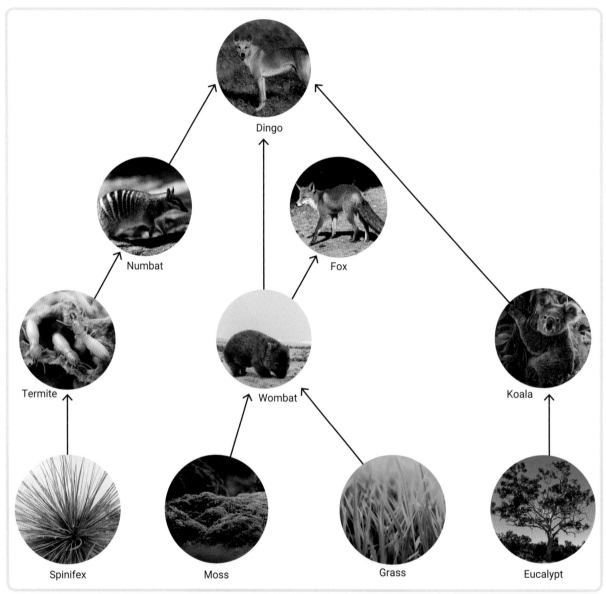

▲ **FIGURE 7.4.2** A food web in an Australian ecosystem

Drawing food webs

Remember that a food web is a series of connected food chains. Therefore, we use arrows to show the direction of energy and matter flow.

Organising a food web can make it easier to interpret. Therefore, if possible, organise organisms according to their trophic levels, with the first trophic level at the bottom and the apex predators at the top.

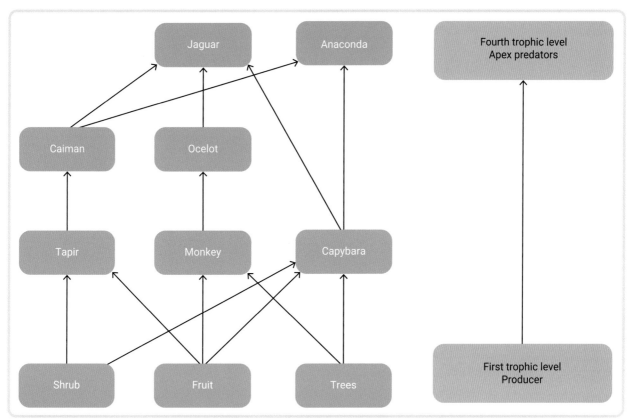

▲ **FIGURE 7.4.3** A food web for a rainforest ecosystem. Note how the organisms have been arranged according to their trophic levels.

Interpreting food webs

Food webs help scientists predict what could happen if one of the organisms in the food web was disrupted in some way. For example, what might happen if a drought caused most of the grass to die in Figure 7.4.2? The wombat would need to eat more moss to survive. If there wasn't enough moss, then the number of wombats in the ecosystem may decrease. They might move to a new area to find food, or they might die of starvation. If there were fewer wombats, then there would be less food for the foxes to eat, reducing their numbers as well. And the dingoes may need to eat more numbats or koalas, affecting their population, too.

> ✩ ACTIVITY

Creating a food web

You need

- pictures of the following organisms with their names

- large piece of butcher's paper and marking pen, or whiteboard, magnets and whiteboard marker

What to do

1. Arrange the organisms into producers and consumers.
2. Place the pictures of the producers at the bottom of the butcher's paper (or whiteboard) to start the food web.
3. Table 7.4.1 lists some food chains for a savanna ecosystem. Use food chain A to place the relevant consumers above the producer. Draw arrows to connect the organisms.

▼ TABLE 7.4.1 Food chains for a savanna ecosystem

A	Red oat grass → warthog → wild dog
B	Acacia → impala → lion
C	Star grass → Thompson's gazelle → Ruppell's vulture
D	Red oat grass → wildebeest → hyena
E	Star grass → hare → lion
F	Star grass → Thompson's gazelle → lion
G	Red oat grass → termite → aardvark → hyena
H	Acacia → impala → cheetah
I	Star grass → warthog → lion
J	Red oat grass → hare → wild dog
K	Star grass → hare → cheetah
L	Red oat grass → wildebeest → lion

4 Repeat step 3 for the other food chains B–L to add the remaining consumers.
  - Remember to try to keep each trophic level on the same level of the food web.
  - You may need to move the organisms (and their arrows) to make them fit in an organised manner.
5 When you have finished, take a photo of your food web or copy it into your notebook.

7.4 LEARNING CHECK

1 **Define** 'food web'.
2 **Explain** why a food web gives a more accurate representation of interactions in an ecosystem than a food chain does.
3 Use Figure 7.4.4 to answer the following questions.

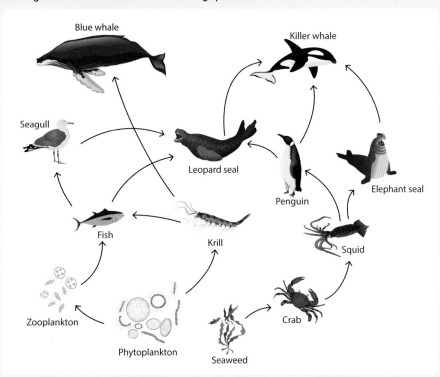

▲ FIGURE 7.4.4 A food web for a marine ecosystem

a **Name** two producers in the food web.
b **Name** an apex predator in the food web.
c **Name** a species that exists in more than one trophic level in the food web.
d **Draw** two separate food chains from this food web.
e Zooplankton and phytoplankton are two organisms in this ecosystem. Research the meaning of their names and **explain** their different roles in the ecosystem.
4 Choose an ecosystem that you are familiar with. **Create** a food web to represent the flow of energy and matter in that ecosystem.

7.5 Movement of energy and matter in an ecosystem

BY THE END OF THIS MODULE, YOU WILL BE ABLE TO:
- ✓ describe the flow of energy in an ecosystem
- ✓ describe the roles of decomposers in an ecosystem
- ✓ explain how matter cycles through an ecosystem
- ✓ compare the movement of energy and matter through an ecosystem.

Quiz
Energy and matter flows

> **GET THINKING**
>
> The movie *Lion King* talks about the 'circle of life'. What do you think this means? Share your thoughts with a partner and discuss this until you come up with an answer that makes sense to both of you.

Both energy and matter move through an ecosystem. However, they do this in slightly different ways. Energy flows through the ecosystem, whereas matter cycles through.

Flow of energy

Energy enters an ecosystem via photosynthesis as light energy, which is converted to chemical energy in sugars. The producers use some of the chemical energy, some is lost as heat and the rest is stored. This means that, when the plant is eaten, only about 10 per cent of the energy remains to be transferred to the consumer.

A similar loss of energy occurs at each trophic level. When something is eaten, approximately 90 per cent of the energy is either used or lost as heat. This leaves only 10 per cent of the energy available for the next trophic level when the organism is consumed. Therefore, at each trophic level, there is less and less energy available. Figure 7.5.1 summarises how energy moved through an ecosystem.

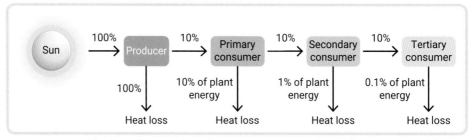

▲ **FIGURE 7.5.1** The flow of energy through an ecosystem. Note how only 10 per cent of the energy at each trophic level is transferred to the next level. This means that only 0.1 per cent of the original energy is available to the tertiary consumer.

Energy enters and leaves an ecosystem. Therefore, we describe its movement as a flow rather than a **cycle**.

cycle
a process that recycles resources

Cycling of matter

decomposer
an organism, such as a fungus or bacteria, that breaks down dead matter

Eventually, all organisms in an ecosystem die. The nutrients and energy from the dead organisms are used by organisms called **decomposers**. Decomposers are organisms, such as bacteria and fungi, that can break down dead organisms into simple nutrients.

Decomposers fulfil an important role in an ecosystem. When they break down the complex structures in plants and animals, nutrients are released in a form that producers can use. Therefore, matter re-enters the food web, allowing it to be continually recycled in an ecosystem.

You can show decomposers in a food web, usually below the producers. However, because they consume the remains of all organisms, there will be many arrows on the food web, which can make it confusing. Therefore, you can leave the decomposers out of the food web. However, we always assume that they are present.

▲ **FIGURE 7.5.2** Decomposers such as mushrooms break down dead organic matter.

Decomposers can be classified as **detritivores** or **saprophytes** depending on how they digest the dead matter. Table 7.5.1 summarises the differences between them.

detritivore
an organism that feeds on dead or decaying matter

saprophyte
an organism that digests dead matter before ingesting it; also known as a saprotroph

▼ **TABLE 7.5.1** A comparison of detritivores and saprophytes

	Detritivore	**Saprophyte**
Method of digestion	Eats dead or decaying matter and then breaks it down	Uses enzymes to break down matter and then eats the nutrients
Where matter is broken down	Internally	Externally
Examples	Worms, beetles	Bacteria, fungi

7.5 LEARNING CHECK

1 **Define**
   a detritivore.
   b saprophyte.
2 **Explain** why energy flows through an ecosystem but matter cycles.
3 **Create** a Venn diagram to show the similarities and differences between the movement of energy and the movement of matter through an ecosystem.
4 **Write** a short story from the viewpoint of a carbon atom that is found in the body of living things and in carbon dioxide in the air. Describe the atom's journey through a food web, and the situations it may find itself in.
5 Do you think it is possible for there to be a very long food chain? **Apply** your understanding of energy flow in an ecosystem to support your answer.

7.6 Modelling ecosystems with pyramids

BY THE END OF THIS MODULE, YOU WILL BE ABLE TO:
- ✓ model the energy, biomass and number of organisms by using pyramids
- ✓ describe, and explain, the shape of energy pyramids, biomass pyramids and number pyramids.

GET THINKING

Look at the pyramids in this module. Can you predict what they are showing you? Why do you think they look like this? Share your answers with a partner – did you both come up with similar answers?

▲ FIGURE 7.6.1 A fire danger sign communicates the risk of fire in a way that everyone can understand.

Using diagrams to organise information

Representing information in a diagram is an efficient way to share data, see relationships and understand the 'big picture'. For example, fire danger signs provide a quick, clear way of communicating the risk of a fire (Figure 7.6.1).

In ecology, information can be organised in different pyramids. Each pyramid is made from a series of bars stacked on top of one another. Each bar reflects the characteristics for a trophic level in a particular food chain, or food web, starting with producers on the bottom. In some pyramids, the edges of the bars are angled to create a smooth, triangular shape.

Energy pyramids

energy pyramid
a graphical representation of the total energy present at each trophic level of an ecosystem

An **energy pyramid** reflects the total energy in each trophic level of a food chain. As energy is lost at each level, the bars become smaller and smaller, creating an upright pyramid.

Interactive resource
Label: Biomass pyramid

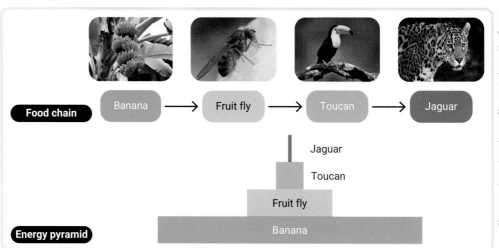

▲ FIGURE 7.6.2 An energy pyramid for a food chain in a rainforest

Biomass pyramids

Biomass is the mass of living things, including plants, animals and micro-organisms. In an ecosystem, the biomass for each population depends on both the mass and number of organisms of a species.

A **biomass pyramid** represents the total biomass for each trophic level of a food chain. The length of the bar represents the amount of biomass: the larger the biomass, the longer the bar. The biomass pyramid for a grassland food chain is shown in Figure 7.6.3.

biomass
the mass of living organisms

biomass pyramid
a graphical representation of the total biomass present at each trophic level of an ecosystem

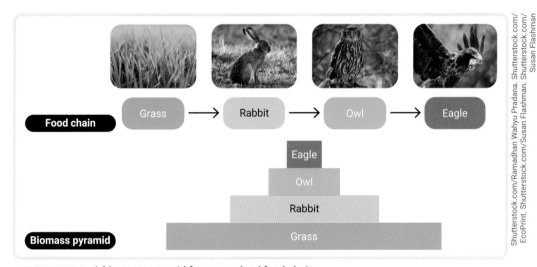

▲ FIGURE 7.6.3 A biomass pyramid for a grassland food chain

In most food chains, the biomass decreases at each trophic level. This is because not all mass is passed onto the next trophic level – it may be lost as waste or may not be consumed.

The exception to this is some marine environments where the pyramid may be inverted, or upside down. This occurs when the organism being eaten reproduces very quickly, replacing the biomass that is lost. For example, phytoplankton is a producer that can grow and reproduce very quickly. Therefore, it can provide food for a population of zooplankton that has a larger biomass than the biomass of phytoplankton. This is shown in Figure 7.6.6 on the next page.

▲ FIGURE 7.6.4 The biomass of eagles is less than the biomass of the owls that they hunt.

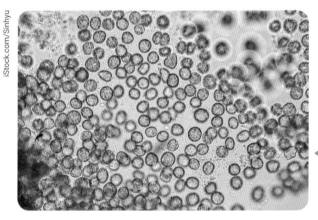

◀ FIGURE 7.6.5 Phytoplankton are microscopic algae. They are producers that can reproduce very quickly, supporting a large biomass of primary consumers.

Chapter 7 | Ecosystems

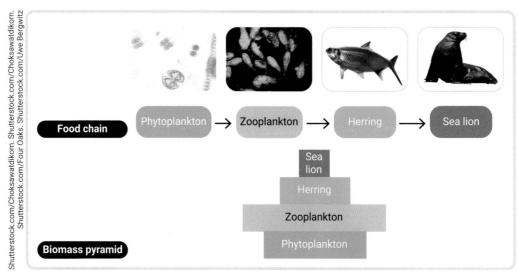

▲ **FIGURE 7.6.6** An inverted pyramid for a food chain in a marine environment.

Numbers pyramids

numbers pyramid
a graphical representation of the total number of organisms at each trophic level of an ecosystem

As its name implies, a **numbers pyramid** represents the number of each organism. In most cases, numbers decrease up a food chain, creating an upright pyramid. For example, in the food chain in Figure 7.6.7, one crested flycatcher would need to eat many grasshoppers to survive.

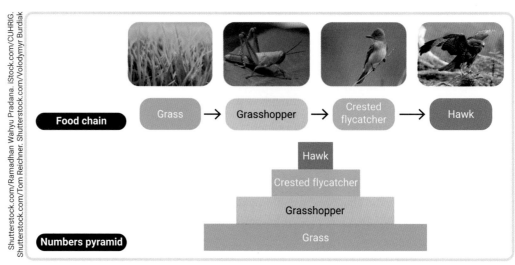

▲ **FIGURE 7.6.7** An upright numbers pyramid

However, numbers pyramids can take any shape because the number of organisms is related to their size. For example, one tree can provide food for a lot of caterpillars, or one dog can provide food for a lot of fleas.

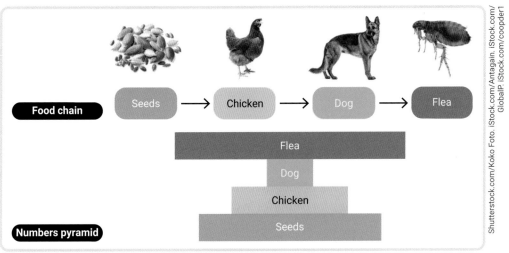

▲ FIGURE 7.6.8 A numbers pyramid with a different shape

▲ FIGURE 7.6.9 Hundreds of fleas can feed on one dog.

7.6 LEARNING CHECK

1. **Define** 'biomass'.
2. **Explain** why energy pyramids are always upright.
3. What type of biomass pyramid may be inverted? **Explain** why.
4. **Draw** an energy pyramid for the following food chain.

   Grass seed → mouse → owl
5. **Create** an energy pyramid, biomass pyramid and numbers pyramid for the following food chain.

   Tree → termite → anteater → jaguar

7.7 First Nations Australians' traditional ecological knowledge

FIRST NATIONS SCIENCE CONTEXTS

IN THIS MODULE YOU WILL:
- explore First Nations Australians' traditional ecological knowledge
- examine the importance of traditional ecological knowledge to restoring ecosystems.

First Nations Australians' traditional ecological knowledge

Over many thousands of years, First Nations Australians have accumulated a deep knowledge of their areas' ecosystems and the natural resources they provide. This is known as traditional ecological knowledge (TEK). It is knowledge that has developed through First Nations Australians' long and sustained contact with their Country/Place, and their experience of living for centuries in close connection with the land.

Seasonal calendars

First Nations Australians have a deep scientific understanding of the complexities and connections between animals, plants, seasonal, weather, and environmental and astronomical changes. This knowledge has been used to devise highly comprehensive seasonal calendars based on thousands of years of observations. These calendars demonstrate understanding of the connection and interactions among and between living things and their environments.

The diversity of climate and ecological zones across Australia means that the seasonal calendars of different First Nations Australians are also quite diverse, with the names and times of the seasons being defined by localised resources and events. For example, the seasonal calendar of the Miriwoong Peoples, whose Country encompasses the east Kimberley region of Western Australia and extends into the Northern Territory, consists of three seasons. In contrast, the seasonal calendar of the D'harawal Peoples of the region north of Sydney consists of six seasons.

Plant and animal seasonal indicators

In traditional times, these calendars allowed First Nations communities to make accurate predictions about recurring seasonal events such as the availability of particular resources or timing of journeys. For example, the First Nations Peoples of D'harawal Country know that the lilly pilly berries start to ripen when they hear the cries of tiger quolls in search of mates. When the lilly pilly berries start to fall, they know it is time to begin their journey to the coast in search of other seasonal resources. The Yawuru Peoples of the Broome region in Western Australia know that it is time to harvest land rather than marine resources by the budding of the bloodwood tree.

▲ FIGURE 7.7.1 Lilly pilly berries. These berries are often associated with significant events within the ecosystem.

Weather seasonal indicators

Annual weather patterns identified by phenomena such as wind strength and direction, appearance of particular cloud type and temperature changes, have long been related to seasonal events. For example, the Yanyuwa Peoples of the Sir Edward Pellow Group of Islands in the Gulf of Carpentaria know that the appearance of 'morning glory' clouds (a rare weather phenomenon that occurs at predictable times) indicate the beginning of the wet season. These clouds also indicate the time when seagulls and sea turtles lay their eggs.

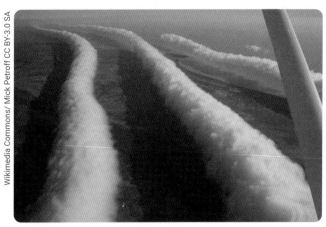

▲ **FIGURE 7.7.2** Morning glory clouds over the Gulf of Carpentaria region mark the beginning of the wet season.

Astronomical seasonal indicators

First Nations Australians' astronomical knowledge, collected through observation over millennia, connects phenomena in the sky with occurrences and events on Earth. For example, the Kaurna Peoples of the Adelaide Plains region in South Australia use the position of the star Parna near the lower left of the moon to mark the beginning of Parnati (autumn), the time when stone fruits are ripe and fish such as whiting are ready for harvesting. The visibility of the constellation Scorpius in the morning sky indicates to the Yirrkala Peoples of east Arnhem Land in the Northern Territory that it is time to trade sea cucumbers.

Integrating seasonal indicators

Around the world, researchers are increasingly recognising and turning to traditional ecological knowledge to provide a more complete understanding of the natural world. The intimate knowledge of seasonal patterns held by First Nations Australians is based on integrating and organising observations from astronomical and weather events, physical changes in plants and the land as well as the presence, movement and habits of animals.

While the Western calendar is based on specific units of time (days, months and seasons), First Nations Australians' calendars are based on cyclical processes in ecosystems – ecological time. First Nations Australians' holistic understandings of their Country/Place allows seemingly separate events to be connected. For example, the Narungga Peoples of Yorke Peninsula in South Australia identify the flowering of the billy button (*Craspedia* species) as a signal that marine resources such as the mulloway are plentiful and, prior to colonisation, indicated it was time to move to coastal areas.

▲ **FIGURE 7.7.3** Billy button flowers signal to the Narungga Peoples that mulloway are plentiful so it is time to move to coastal areas.

Recognising traditional ecological knowledge

The importance of including First Nations Australians and their traditional ecological knowledge and viewpoints in natural resource management is being increasingly recognised.

First Nations Australians sustainably managed Australian ecosystems for thousands of years prior to colonisation. As a result, learning from them should contribute to more sustainable resource use, conservation of biodiversity and natural ecosystem restoration.

Traditional Owners are increasingly working with scientific and government organisations to inform land management practices and restoration processes in areas where the impact of colonisation and introduced species has damaged environments.

▲ FIGURE 7.7.4 (a) Cane toads and (b) gamba grass are introduced species that have had significant negative impacts on Australian ecosystems.

ACTIVITY

1. **Explain** what is meant by traditional ecological knowledge.
2. **Suggest** why First Nations Australian groups have different calendars.
3. **Describe** the advantages of using First Nations Australians' seasonal calendars to understand ecosystems.
4. **Propose** why traditional ecological knowledge should be used when assessing and repairing damaged ecosystems.
5. **Discuss** how First Nations Australians' seasonal calendars and traditional ecological knowledge can be used to assess the effect of the changing climate on ecosystems.

7.8 Human impact – introduced species

SCIENCE AS A HUMAN ENDEAVOUR

BY THE END OF THIS MODULE, YOU WILL BE ABLE TO:
- ✓ define native species, invasive species and endangered species
- ✓ explain how an event affects an ecosystem.

Introduced species in Australia

As humans, we have a responsibility to consider the impact of our actions on the environment. Unfortunately, humans have made changes that have had negative impacts on ecosystems, including climate change, **deforestation**, pollution, poaching and mining.

Another change that can have negative impacts is the introduction of species that are not native to an ecosystem. When Australia was colonised, food plants such as wheat and corn, and animals such as cows, sheep and goats were introduced. To create gardens, roses, blackberry bushes and ivy were planted. Rabbits, foxes and European carp were released into the wild for sport. Unfortunately, there was no understanding of the consequences of these actions on the **native species** of Australia.

These plants and animals are called **introduced species** because they have either arrived accidentally or been deliberately brought to an area. Sometimes the local conditions are not favourable to the introduced species, and it struggles to survive. However, some introduced species thrive in their new environment because they are free from the predators and diseases that kept their numbers in check in their native environment. These species cause harm to their new environment and are termed **invasive species**.

If the new environmental conditions are favourable, introduced species will reproduce quickly, and their population size will increase. As their numbers increase, they compete with native species for food and shelter. This can cause a decrease in the population of native species, which may become **endangered** or even **extinct**.

deforestation
the removal of naturally occurring forest by logging or burning

native species
an organism that originated and developed in the environment

introduced species
a species that was not part of the original ecosystem; for example, plants, animals and micro-organisms brought into Australia from other countries

invasive species
an introduced species that disrupts the ecosystem

endangered
in danger of becoming extinct

extinct
no longer in existence

Controlling introduced species

Overall, the impact of introduced species can be extremely detrimental to a local ecosystem. Tables 7.8.1 and 7.8.2 give examples of deliberate and accidental species introductions to Australia. Perhaps the most well-known example is the rabbit. Rabbits arrived with the First Fleet in 1788 and were kept as pets and as a source of food. However, it was the deliberate introduction in 1859 of rabbits for sport that probably led to a devastating rabbit plague in 1890. The rabbit population reached 600 million before it was partly controlled by the release in 1950 of the myxoma virus, which causes the disease myxomatosis that kills rabbits.

▲ FIGURE 7.8.1 Rabbits reached plague numbers before the introduction of biological control.

▲ FIGURE 7.8.2 Prickly pear is an introduced species to Australia.

Another example is the prickly pear, which was brought to Australia on the First Fleet. It infested 25 million hectares of land in New South Wales and Queensland. This was eventually controlled by releasing a caterpillar that ate only prickly pear.

The caterpillar that ate the prickly pear and the myxoma virus that killed rabbits are two examples of a **biological control**.

In north Queensland, the native cane beetle was destroying sugar cane crops. In 1935, the cane toad was deliberately introduced from South America as a method of biological control. However, the cane toad did not reduce the population of cane beetles. Instead, it adjusted quickly to the tropical environment, eating almost anything it could swallow and spreading disease. Birds and animals that tried to eat the cane toad were poisoned. The rapid increase in cane toad numbers led to the decline of native frogs and reptiles in Queensland. Since then, cane toads have spread through Queensland and into New South Wales. They reached the Northern Territory in 1974 and Western Australia in 2009.

biological control
the reduction of a pest species by using natural enemies

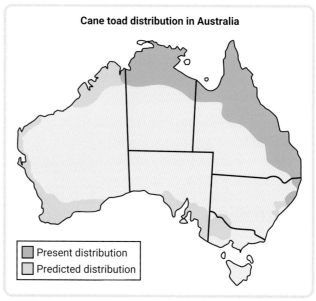

▲ FIGURE 7.8.3 The cane toad is an introduced species that has spread from New South Wales across to Western Australia as a result of a lack of natural predators.

▼ TABLE 7.8.1 Intentionally introduced species to Australia

Species	Year of introduction	Reason for introduction
Foxes	1855	Fur trade, sport
European carp	1859	Food source, sport
Rabbits	1859	Sport
Cane toad	1935	Biological control of the cane beetle

▼ TABLE 7.8.2 Accidentally introduced species to Australia

Species	Location	How it happened
House mouse	Across Australia	Arrived on boats from England to Australia
Red-eared slider tortoise	New South Wales	Escaped from household pet aquariums
Pacific starfish	Victoria	Dumped with ballast water from international ships

7.8 LEARNING CHECK

1. **List** three species that have been introduced to Australia.
2. **Explain** why a lack of natural predators leads to an introduced species disrupting the ecosystem.
3. **Compare** invasive species and introduced species.
4. In 2019, the Australian Government provided $12 million towards controlling cane toads. **Explain** why cane toads pose a threat large enough to justify this spending.
5. The rabbit-proof fence is an initiative that the Western Australian Government implemented in 1907 to protect the state. **Discuss** the use of the fence to keep rabbits out of Western Australia. Consider why it was thought to be necessary, the advantages of this strategy over other methods, and reasons that the fence may not succeed.

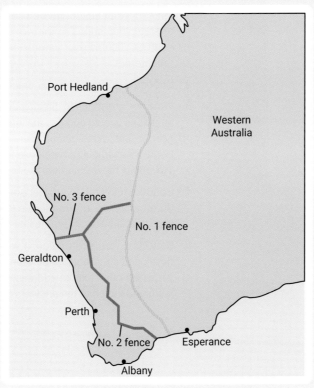

▲ FIGURE 7.8.4 The locations of rabbit-proof fences in Western Australia

6. Find out what species have been introduced to the area where you live. What impact have they had on the environment?

Weblink
Rabbit-proof fence

Video activity
Cane toads

SCIENCE INVESTIGATIONS

7.9 Science communication

SCIENCE SKILLS IN FOCUS

IN THIS MODULE, YOU WILL FOCUS ON LEARNING AND IMPROVING THESE SKILLS:

- investigating a local, national or global environmental issue
- communicating scientific information clearly and accurately via a format that is appropriate to the audience.

Communicating means to share information. To communicate effectively, it is important to consider how we are sharing the information so that the message is received. Figure 7.9.1 summarises key factors for communicating effectively.

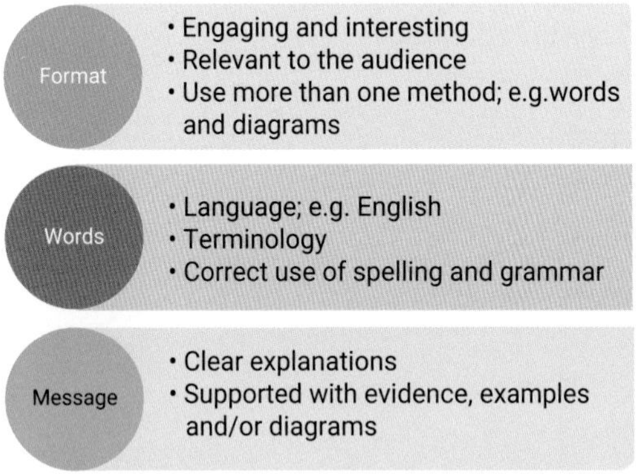

Format
- Engaging and interesting
- Relevant to the audience
- Use more than one method; e.g. words and diagrams

Words
- Language; e.g. English
- Terminology
- Correct use of spelling and grammar

Message
- Clear explanations
- Supported with evidence, examples and/or diagrams

▲ FIGURE 7.9.1 Factors to consider for effective communication

Video
Science skills in a minute: Science communication

Science skills resource
Science skills in practice: Science communication

UNDERSTANDING AN ENVIRONMENTAL ISSUE

AIM

To investigate an environmental issue to share an understanding about:
- what the problem is
- what caused the problem
- what is being, or has been done, to control the problem

YOU NEED

☑ access to the Internet

WHAT TO DO

1. Find out what environmental issues affect your area or state, the country or the world. Use a variety of sources such as your parents, teachers, library and the Internet.
2. Choose one issue that you are interested in.
3. Write five questions to direct your research. These questions should address the information that you need to find to understand your chosen environmental issue; for example, 'Where does the issue occur?' or 'How does the issue affect the food web?'
4. Conduct research to answer your questions. Summarise the information, in your own words, noting its source, in a table like Table 7.9.1.
5. Reflect on your information. If you have not developed a thorough understanding of the issue, write new questions to address what you still need to learn and then find the answers to these questions.

RESULTS

Use Table 7.9.1 to summarise your information for your first question. Create a new table for each research question.

▼ TABLE 7.9.1 Notes for environmental issue research

Question 1	
Summary of information	**Source of information**

PUTTING IT TOGETHER

1. Choose how you would like to use the information about the environmental issue. For example, you may educate others about the issue, inspire others to make individual changes, fundraise to put solutions into action, or advocate for changes to laws and standards.

2. Choose an audience with whom to share what you have learned. This may be:
   - other students
   - young children
   - local, state or federal government representatives
   - your school
   - the broader community.

3. Choose a mode of communication that will allow you to engage with your audience to achieve your goal. For example, a children's book would be suitable for young children, a letter to your local politician would address the government, or a newspaper article would reach the broader community. Other ideas are a website, an infographic, a poster, a fundraising event, a video, a podcast or a short story.

4. Use your notes to explain your chosen environmental issue to your chosen audience.

5. Collate the information about your sources in a bibliography. Your teacher will tell you what format to use; however, remember to include all relevant information such as the author, title, website name, URL and access date, and to list your sources in alphabetical order based on the author's surname.

7 REVIEW

REMEMBERING

1 **Name** five biotic and five abiotic factors in a marine ecosystem.
2 **Describe** the process of photosynthesis, including the substances that are used and produced.
3 **Write** the word equation for cellular respiration.
4 **Describe** the difference between an endangered species and an extinct species.
5 **Define:**
   a   producer.
   b   consumer.
   c   apex predator.
   d   decomposer.
6 **Describe** the flow of energy through an ecosystem.

UNDERSTANDING

7 **Explain** the difference between an environment and an ecosystem.
8 **Explain** why producers are always at the bottom of a food chain or web.
9 Why do organisms need cellular respiration?
10 **Explain** the difference between a food chain and a food web.
11 Why is photosynthesis such an important process? **Explain** your answer in terms of energy.

APPLYING

12 **Describe** your school environment.
13 Draw a food chain that includes:
   a   cow, grass, human.
   b   small bird, pollen, spider, bee.
   c   fox, eucalyptus tree, koala, eagle.
14 The following diagram shows food chain for a marine environment.
   a   **Identify** the producer and the primary, secondary and tertiary consumers.
   b   **Draw:**
      i   an energy pyramid for the food chain.
      ii a biomass pyramid for the food chain.

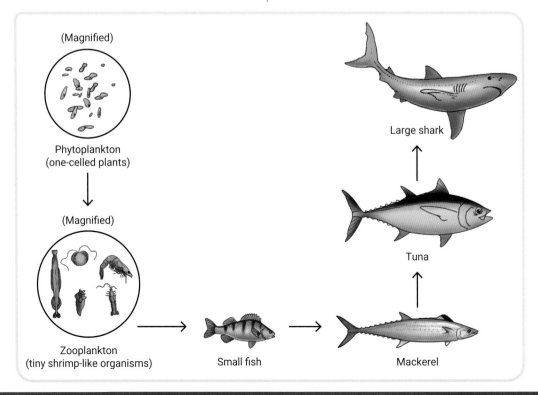

15 Use the following diagram of a typical Australian food web to answer the questions.

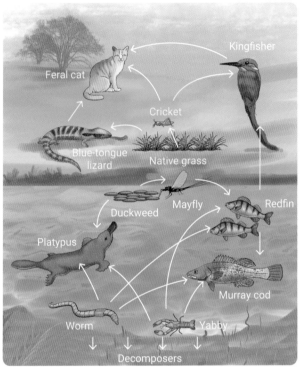

a **State** which organisms are producers and which are consumers.
b Give an **example** of a herbivore, an omnivore and a carnivore.
c **Explain** the difference between an autotroph and a heterotroph, and give an example of each from the food web.
d **Write** three food chains that are found in this food web.
e **Explain** the effect the feral cat has on the food web.
f **Explain** what might happen if the European rabbit was introduced into this community.

16 First Nations Australians' seasonal calendars use a range of seasonal indicators. **Describe** what indicators in your local environment could be used to develop your own seasonal calendar.

EVALUATING

17 **Discuss** the advantages and disadvantages of intentionally releasing introduced species into Australia.

18 Extension: **Construct** a food chain that could be represented in each of the following pyramids.
   a Energy

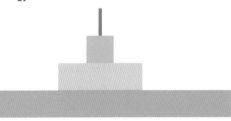

   b Biomass

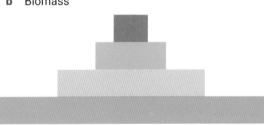

   c Numbers

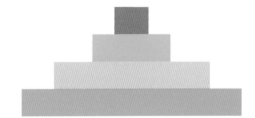

19 Extension: Is the following energy pyramid drawn to scale? **Explain** your answer.

20 With the introduction of the European rabbit and fox into Australia, the population of native bilbies declined dramatically. What conservation technique do you think would help increase the numbers of the bilbies? **Justify** your answer.

21 Feral cats have become one of the greatest threats to the survival of many native species.
   a **Explain** how a feral cat could affect an ecosystem.
   b **Suggest** a method of solving this problem and **explain** why it would work.

CREATING

22 **Write** a sentence that uses each of the following groups of words.
   a Producer, consumer, apex predator
   b Producer, herbivore
   c Food chain, food web, ecosystem
   d Autotroph, heterotroph, producer, consumer
   e Decomposer, matter, cycle, producer

23 **Create** a flow chart to model how matter cycles through an ecosystem.

24 Find three species that have not been mentioned in this chapter, that have been introduced into Australia. How have these species affected the environment?

25 **Create** a food chain for the following information.

One day, a bird saw a snake with a full belly, lazing in the late-afternoon sunshine. The bird thought, 'Mmm, I'm hungry.' It flew down and caught the snake for dinner. Earlier that day, the snake had been busy catching its own meal. Gliding by the pond, it had noticed a grasshopper nibbling on some grass. When a cheeky frog snapped up the grasshopper with its long, sticky tongue, the snake thought, 'Jackpot!' It slithered quickly to the edge of the pond and caught the frog as it was eating the grasshopper.

BIG SCIENCE CHALLENGE PROJECT #7

Wooroloo, in Western Australia, is an area of beautiful bushland. On 1 February 2021, a devastating bush fire started. The fire lasted 6 days, destroying 86 homes and 10 900 hectares of land. The photos are just a few images from the fire.

▲ The Wooroloo bushland

1 Connect what you've learned

At the start of this chapter, you summarised how you thought the plants and animals would have been affected during, immediately after, and one year after the fires. Add to your summary by including information about how each of the following would be affected at different times.

- Abiotic factors
- Biotic factors
- Food chains

2 Check your thinking

Choose one plant or animal species that is native to your local area. Brainstorm information about the ecosystem that it lives in, including the abiotic factors and biotic factors that it needs to survive.

Imagine that a fire destroyed the area. What would your organism need to survive in the environment again? How could these needs be met? How could the risk of a fire in the future be reduced?

3 Make an action plan

Use your answers from step 2 to develop a plan to protect the native species that you chose. In your plan, consider what needs to be done, who will do it, what resources you will need and how much it will cost.

4 Communicate

Create a radio, television or online advertisement to gain support for your action plan.

▲ The impact of the Wooroloo bushfire in February 2021: the fire, the burned land and the injured animals

8 Forces

8.1 What is a force? (p. 220)
A force is a push, pull, twist or squeeze applied by an object to another object.

8.2 Measuring forces (p. 224)
Forces can be measured by a device such as a spring balance. The unit used to describe the size of the force is the newton.

8.3 Contact forces (p. 226)
Contact forces are exerted on objects by other objects that are touching them or connected to them via a string, cord or cable.

8.4 Non-contact forces (p. 228)
Some objects can exert a non-contact force at a distance without any contact or connection.

8.5 Balanced and unbalanced forces (p. 232)
Balanced forces occur when the individual forces on an object cancel each other out. Unbalanced forces occur when the individual forces do not cancel each other out, so there is an overall force on an object.

8.7 The effect of mass (p. 236)
When an object experiences a force, its mass plays a role in determining the resulting motion.

8.6 Net force (p. 234)
The sum of two or more forces is called the net force.

8.8 Mass and weight (p. 238)
The weight of an object depends on gravity and mass. The weight of an object can change if mass changes or if the object is moved to a place with different gravity.

8.9 Simple machines (p. 242)
A simple machine makes a task easier by multiplying the force applied.

8.10 FIRST NATIONS SCIENCE CONTEXTS: First Nations Australians' spear-throwing technology (p. 248)
First Nations Australians developed and produced tools such as spear-throwers, which increase the speed and accuracy of hunting spears.

8.11 SCIENCE AS A HUMAN ENDEAVOUR: Principles of rocket propulsion (p. 249)
All rockets use the simple principles of physics to achieve extraordinary outcomes.

8.12 SCIENCE INVESTIGATIONS: Analysing data in tables and graphs (p. 250)
The relationship between weight and friction

BIG SCIENCE CHALLENGE #8

▲ FIGURE 8.0.1 These very heavy statues were moved several kilometres.

The photo shows some of the statues on Easter Island. It is estimated that hundreds of statues were moved into position between 650 and 400 years ago. Although they vary in size, their average mass is approximately 55 tonnes (55 000 kg) and they are about 7 metres high. Each statue has been moved several kilometres over hilly ground from where they were carved from volcanic rock.

▶ How could an object so huge be moved by people without sophisticated machinery?

▶ How can we use a knowledge of forces to overcome the challenge to lift and move heavy objects over a distance?

#8 SCIENCE CHALLENGE ACCEPTED!

At the end of this chapter, you can complete Big Science Challenge Project #8. You can use the information you learn in this chapter to complete the project.

Assessments
- Prior knowledge quiz
- Chapter review questions
- End-of-chapter test
- Portfolio assessment task: Science investigation

Videos
- Science skills in a minute: Representing data **(8.12)**
- Video activities: Friction **(8.3)**; What are magnets? **(8.4)**; Measuring weight in space **(8.8)**; Levers, wheels and pulleys **(8.9)**; Rockets 101 **(8.11)**

Science skills resources
- Science skills in practice: Collecting and representing data **(8.12)**
- Extra science investigations: The effect of friction **(8.3)**; Exploring electrostatic forces **(8.4)**; Tug-o-war forces **(8.5)**

Interactive resources
- Simulations: Effect of friction **(8.3)**; Find the net force **(8.6)**; How does mass affect force? **(8.7)**
- Label: Applied and reaction forces **(8.1)**
- Match: Simple machines **(8.9)**
- Drag and drop: Classifying contact and non-contact forces **(8.4)**

To access these resources and many more, visit
cengage.com.au/nelsonmindtap

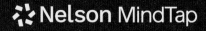

Chapter 8 | Forces

8.1 What is a force?

BY THE END OF THIS MODULE, YOU WILL BE ABLE TO:
- ✓ list examples of forces
- ✓ list the effects of a force.

Interactive resource
Label: Applied and reaction forces

GET THINKING

Why do some things (such as an empty box or a ball) move when we bump into them? Why do other objects (such as a piano or a car) not move when we bump into them? Can you think of a key difference between these groups?

What are forces and where do they come from?

Look around at the objects you can observe. Whether those objects are in **motion** or not moving, each is doing so as a result of forces. We cannot see **forces**, but we can often observe the result of them.

Pushes, pulls, twists and squeezes can all be described as forces. Pushes, pulls, twists and squeezes are all applied by one object on another object. You can see examples of pushes, pulls, twists and squeezes in Figures 8.1.1–8.1.4.

motion
the change in position of an object over time

force
a push, pull, twist or squeeze experienced by an object when it interacts with another object

▲ FIGURE 8.1.1 By pushing the box, the child is applying a force to it.

▲ FIGURE 8.1.2 By pulling on the lead, the person is applying a force on the dog.

▲ FIGURE 8.1.3 A twist is a force.

▲ FIGURE 8.1.4 A squeeze is a force.

Therefore, forces are exerted by one object on another object. This can also be described as an **interaction** between the two objects. For example, a racquet exerts a force on a ball in a tennis game; therefore, there is an interaction between the ball and the racquet (Figure 8.1.5). A wall, the ground and **gravity** each exerts a force on a ladder leaning against a wall (Figure 8.1.6).

interaction
an action that occurs as two objects have an effect on each other

gravity
a force applied by one mass on another mass

▲ FIGURE 8.1.5 When a tennis racquet exerts a force on a ball, there is an interaction between the ball and the racquet.

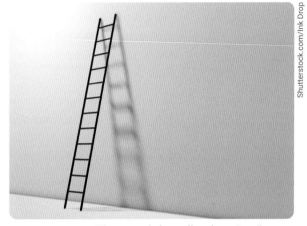

▲ FIGURE 8.1.6 The ground, the wall and gravity all exert a force on the ladder.

What do forces do and where do they come from?

Forces can hold an object in place such as those that are acting on the ladder in Figure 8.1.6.

A force applied to an object can result in a change in the object's shape or motion. Changes in motion include a change in:

- direction, such as when an object moves around a corner or when it bounces off a surface
- speed, such as when an object goes faster or slower.

Often, a change of direction and a change of speed both occur to an object as a result of the action of a force.

Changes of shape such as squashing a rubber ball or bending a paperclip can be complicated and will not be explored in this chapter.

▲ FIGURE 8.1.7 The shot-putter exerts a force on the shot (large metal ball) to change its direction and speed.

Forces come in pairs

Whenever you can identify one force acting, there must be another force acting. These forces are exactly the same size but opposite in direction and each one acts on one of the two objects in the interaction. If the force you have identified can be described in the form 'object A exerts a force on object B', then it must be true that 'object B exerts an equal force in the opposite direction on object A'. One, usually applied by a person, animal or machine, is described as an **applied force**. The other is described as a **reaction force**. The applied force acts on one object involved in the interaction and the reaction force acts on the other object.

For example, as a sprinter starts a race, their feet apply a backwards force on the blocks (applied force). The block applies an equal and opposite force forwards on the sprinter's feet (reaction force). This reaction force is the force that causes the sprinter to increase speed at the start of the race.

applied force
a force that is applied to an object by another object

reaction force
a force acting in the opposite direction to the applied force and on the object that exerted the applied force

▲ FIGURE 8.1.8   The sprinter pushes back on the blocks; the blocks push forward on the sprinter.

A simple phrase can provide a structure that you can use whenever you are looking at force pairs. For example, 'Sprinter pushes back on blocks; blocks push forwards on sprinter'. Often the second force of a pair is not obvious, but it is always there.

Extension challenge: Tug of war

ACTIVITY 8.1

You will probably have watched or taken part in a tug-of-war. A tug-of-war is a challenge of forces.

Imagine a simple tug-of-war with only one person on each end of the rope. Use the information you have learned about forces in this module to describe all the forces involved in this tug-of-war. A good way of doing this is to draw a simplified diagram of the two people standing on the ground with a rope between them, as shown in Figure 8.1.9, and then draw labelled arrows to represent the forces.

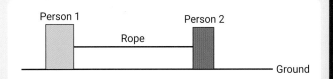

▲ FIGURE 8.1.9 A simple representation of two people playing tug-of-war

Recall that forces are an interaction between two objects and that each force should be part of a pair of forces, each acting on one of the two objects in the interaction. The forces on each person should be the mirror of each other. How many forces have you identified and labelled on the diagram? If you have four pairs of forces – two pairs associated with each person, then you are on the right track! Check with the person next to you. When both of you are satisfied, check against your teacher's example.

In terms of the correct diagram of forces in the tug-of-war – how does one person win? An immediate answer might have come to mind but try considering a slight change to the tug-of-war set-up. Imagine person 1 is standing on slippery ground whereas person 2 is wearing boots with studs so that they have an excellent grip. It is clear that person 2 will win, but in terms of the forces in the tug-of-war, what has changed? Are you now able to give a better answer as to how one person can win this tug-of-war?

8.1 LEARNING CHECK

1. What is a force?
2. **List** three effects a force might have on an object.
3. Imagine a book on a shelf. What forces are acting on the book and in which direction?
4. In one minute, **list** as many objects as you can that are *not* experiencing forces. **Compare** your list with that of a partner. Did you agree on any?
5. **Describe** the interactions needed to lift a pen off the desk in terms of forces applied by an object to an object.
6. **Identify** an applied force–reaction force pair you can see or think of. Write it in the form of 'Sprinter pushes back on blocks; blocks push forwards on sprinter'.
7. **Classify** the following forces as a pull, a push, a twist or a squeeze.
    a. Turning a door handle
    b. Sealing a ziplock bag
    c. Opening a drawer
    d. Sliding a calculator away from you across the desk

8.2 Measuring forces

BY THE END OF THIS MODULE, YOU WILL BE ABLE TO:
✓ describe how to use a device to measure force.

Quiz
Who was Newton?

GET THINKING

How do you explain to someone how much longer your pencil is than your eraser, or how much hotter today is than yesterday? You could do a qualitative comparison, which uses words to compare, or a quantitative comparison, which uses numbers to compare. What would you need to make a quantitative comparison? See if you have worked it out by the end of the module.

Describing force

unit
a fixed quantity used as a standard of measurement

Whenever we describe a quantity, we use a number and a **unit**. The number gives information about the quantity and the unit indicates the scale of measurement. We might describe a quantity by saying 'she jumped 3.4 metres' or 'he drank 200 millilitres' or 'I spent $12.50'. Each of these statements uses a number and a unit. If either the number or unit is missing, the information is incomplete and unclear. Similarly, when we describe forces, we need to use a number and a unit.

newton
the unit of force (N)

Force is measured in units called **newtons**, which are often abbreviated to the capital letter N. For example, in order to lift a medium-sized book from the table, you might need to exert an upwards force on the book of 3–4 N. This unit is named after scientist Sir Isaac Newton, whose contributions in the 17th century enabled great progress in our understanding of forces.

Measuring force

Forces can be measured with a device called a force meter. It is also called a newton meter or **spring balance**. It has hooks at each end – one that is fixed and the other on a spring (Figure 8.2.1). When the fixed hook is held in place and a force is applied to the other hook, the spring stretches. The distance that the spring stretches is related to the size of the force. This means that when the force doubles, the amount the spring stretches doubles, too. As the spring stretches, it moves a pointer on the device. The force value, in newtons, can be read from the position of the pointer against the scale.

spring balance
a device for measuring force; also called a force meter or newton meter

There are also digital force gauges that give a digital value of the size of a push or a pull.

▲ FIGURE 8.2.1 A spring balance measuring the downwards force of a bag of potatoes

Scales research project

ACTIVITY 8.2

▲ FIGURE 8.2.2 Some examples of different types of scales

Perhaps you have stepped onto a set of scales to find out how heavy you are, or you have seen someone else do it. Alternatively, you might have used or seen used a set of kitchen scales. Most commonly, the scales will have a digital display or a scale with lines and numbers (analogue) that moves past a pointer until it settles in a particular place.

You may have seen scales that feature two pans on either side of a central column, to which the two pans are connected by a rod that rests on the top of the column. This is called a balance scale.

You can see examples of scales in Figure 8.2.2. The scales are used to obtain a measurement in kilograms (kg) or grams (g), which are the units of mass.

It might surprise you to learn that mass is actually quite tricky to measure directly. So, what do these scales actually measure and how?

1 Research what the analogue and digital scales measure and how this measurement is used to give information about the mass of the object on the scale.
2 Research how forces are important in the function of the balance scale. Can you think of a way a balance scale might measure the wrong value?

8.2 LEARNING CHECK

1 **State** the unit of force.
2 **Research** Sir Isaac Newton and make a list of three of his contributions to science and mathematics.
3 **Explain** why springs and rubber bands are good at measuring forces.
4 **Justify** the importance of having a unit for force that is agreed on by everyone.
5 **Describe** how you would make your own force meter from a rubber band, a ruler, a piece of cardboard and some tape. How could you make sure the measured values were accurate? Have a class challenge to see who can make the most accurate force meter from just this equipment.

8.3 Contact forces

BY THE END OF THIS MODULE, YOU WILL BE ABLE TO:
- ✓ describe a range of contact forces
- ✓ describe the possible effects of a contact force on an object.

contact force
a force applied by one object to another object when they are touching each other

pushing force
a force applied by an object towards another object

pulling force
a force applied by an object away from another object

tension force
a force that acts to pull along a rope, cable, string, wire or chain

friction
a force that acts against the direction of motion, or intended motion, of an object because of an interaction between their surfaces

GET THINKING

Make a list of all the things touching you right now. Can you feel or predict the direction of the force they are exerting on you, even if it is very small? Sketch a diagram to represent the forces.

Contact forces

Many of the forces that we witness, experience or exert are **contact forces**. A contact force is one where the object exerting the force and the object experiencing the force are in contact with (touching) each other. For example, when your finger pushes a calculator button or your hand pulls down a blind, this is a contact force. Each interaction involves objects that are connected.

Analysing contact forces

There are several types of contact forces. Any time you can see the surface of one object touching the surface of another object, there will be a contact **pushing force**. Your finger pushing on a calculator button is an example of a contact pushing force.

Contact **pulling forces** are those that act to move an object away from or past another object, such as the pulling force you apply to lift your school bag from the floor or to drag the garbage bin past the door. Often, contact pulling forces act along cables, strings, cords, chains or wires. The cord used to pull down a blind is exerting a contact pulling force.

Contact pulling forces that act along things such as a cable, string, cord, chain or wire are described as **tension forces**. We might say that, in Figure 8.3.2, the tension in the cord pulling down the blind is 12 N.

Friction

Friction is a force that resists the movement of one object against another. It is a contact force that acts against the direction of motion of an object or intended motion of an object.

▲ FIGURE 8.3.1 A finger exerts a contact force on a calculator button.

▲ FIGURE 8.3.2 A cord exerts a contact force pulling down on the blind.

If an object is on a surface, there will be friction because of the texture of the two surfaces at a microscopic level. Even very smooth surfaces are covered completely with pits and bumps. When two surfaces are in contact, some of the pits of one inevitably fit with some of the bumps of the other. This causes friction, which prevents, or slows, motion.

Friction makes it harder to move things. This might seem like a problem, but without friction you couldn't walk, hold a pen, or manage many other actions. We rely on friction between our feet or shoes and the ground so that we don't slip and so we can push ourselves forward. We rely on the friction between our fingers and the pen so that we can maintain its position in our fingers while we write.

▲ **FIGURE 8.3.3** Even smooth surfaces are covered in pits and bumps that result in friction.

Air resistance

Air resistance is a specific form of friction. When an object such as a train is moving forward, the front of the train interacts with particles in the air. Each air particle exerts a tiny force on the train against the direction of the train's travel. The vast number of particles means that the total force opposing the motion is significant. In addition, the sides of the train move past the air. Irregularities (pits and bumps) on the side of the train provide additional opportunities for interactions with air particles, which will exert additional backwards forces.

Air resistance is friction that only occurs when objects are moving. However, friction is also present when objects are stationary. For example, if you were trying to move a heavy set of bookshelves across a carpeted floor, you might push very hard without causing any movement. Friction between the bookshelves and the floor would be the force that is acting to prevent the movement.

air resistance
friction caused by an object's motion through the air

Video activity
Friction

Interactive resource
Simulation: Effect of friction

Extra science investigation
The effect of friction

8.3 LEARNING CHECK

1 **List** the types of contact forces.
2 What is tension? Use an example in your response.
3 **Describe** why there would be friction between the sole of your shoe and the road.
4 Why does air resistance increase as you move faster?
5 **Identify** three applied contact force–reaction force pairs you can see or think of. Write them in the form of 'sprinter pushes back on blocks; blocks push forwards on sprinter'.
6 **Classify** the following situations as a push, a pull, friction, air resistance or a combination.
  a A skydiver suddenly slows as their parachute opens.
  b An elevator moves up a building.
  c A chair is harder to move when someone is sitting in it.
  d An athlete raises a weight from shoulder height to above their head.

8.4 Non-contact forces

BY THE END OF THIS MODULE, YOU WILL BE ABLE TO:
- ✓ describe the key non-contact forces
- ✓ describe situations in which non-contact forces are observed.

> **GET THINKING**
>
> Why does Earth stay in orbit around the Sun? How can the Sun influence Earth when it is so far away? Discuss this with a partner and come up with a reason that you both agree on.

Non-contact forces and force fields

non-contact force
a force that an object exerts on another object without the objects touching each other

field
a region of space in which a non-contact force exists

A **non-contact force** is a force that one object can exert on another object without the objects needing to touch or be connected. Non-contact forces can, therefore, act from a distance.

When one object (A) exerts a force on another object (B) without contact, we say that object A is surrounded by a **field** known as a force field. A field is a place where an object, such as object B, experiences a non-contact force.

There are three types of non-contact force. Two of the types of non-contact forces can be attractive or repulsive, depending on the circumstances, as you will see later in this module. However, the third type of non-contact force is always attractive.

Types of non-contact forces

gravitational force
a force that results from the mass of an object and acts on other objects

magnetic force
a force acting between magnetic poles

electrostatic force
a force resulting from the electrical charge of two objects

The three types of non-contact forces are **gravitational force**, **magnetic force** and **electrostatic force**. You will have probably have seen, and experienced, each of these many times.

Gravitational forces are non-contact forces that act between all objects, attracting one object to another. Gravitational forces are relatively weak; however, their strength depends on the mass of the two interacting objects. The greater the mass, the stronger the force. We only notice gravitational forces when they are strong enough because one of the objects is very heavy. That is why we don't notice the gravitational force between our body and a pen, but we do notice the gravitational force between our body and Earth.

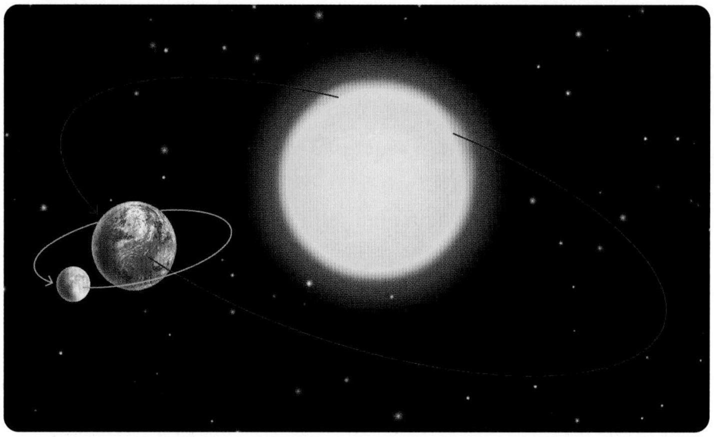

▲ **FIGURE 8.4.1** Earth is attracted to the Sun by a gravitational force and the Moon is attracted to Earth by a gravitational force.

Magnetic forces are non-contact forces that act between two objects that are magnetic in nature. This could be two **magnets** or a magnet and an object that can become temporarily magnetic, such as a paperclip. Magnets can be made of a few common metals, such as iron, cobalt and nickel, or of rarer metals, such as neodymium. A magnet or a temporarily magnetic object consists of two opposite magnetic poles – called the north and south poles. The unlike poles of two magnets exert an attractive force on each other, so a north pole attracts a south pole of another magnet. Two like poles exert a repulsive force on each other, so any two adjacent north poles will repel and any two adjacent south poles will repel.

magnet
a material that produces a magnetic field

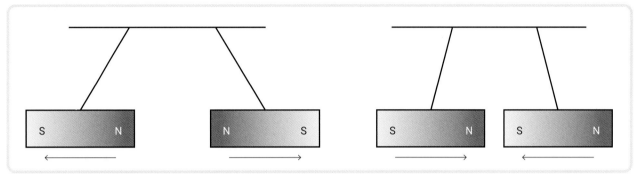

▲ **FIGURE 8.4.2** Two north poles of adjacent magnets will repel each other. One south pole and one north pole of adjacent magnets exert an attractive magnetic force on each other.

Electrostatic forces are non-contact forces that act between two objects that have an electrical charge. This force is often observed when we use friction to make an object charged, such as rubbing a balloon on our hair. The friction results in the balloon becoming negatively charged and your hair becoming positively charged. The balloon and your hair then exert an equal and opposite force on each other that causes them to be attracted to one another.

Video activity
What are magnets?

Interactive resource
Drag and drop: Classifying contact and non-contact forces

Extra science investigation
Exploring electrostatic forces

▲ **FIGURE 8.4.3** The boy's hair is attracted to the balloon by an electrostatic force.

Features of non-contact forces

Objects that have mass are described as being surrounded by a gravitational field. You have a gravitational field, but it is very weak because your mass is small. Earth has a gravitational field that is much, much stronger and so the non-contact gravitational forces it exerts are greater and more noticeable.

Objects that are magnetic are described as being surrounded by a magnetic field. If another magnetic object is placed in this magnetic field, it will experience a non-contact magnetic force.

Objects that are electrically charged are described as being surrounded by an electric field. If another electrically charged (positive or negative) object is placed in this electric field, it will experience a non-contact electrostatic force.

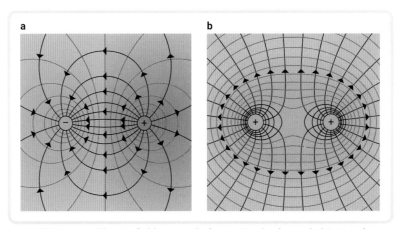

▲ **FIGURE 8.4.4** Electric fields around (a) a positively charged object and a negatively charged object, and (b) two positively charged objects. The intensity of colour indicates the strength of the field. The arrows show the direction of force on a positive charge in the field.

Effect of non-contact forces

attractive force
a non-contact force that brings two objects closer together

repulsive force
a non-contact force that pushes two objects away from each other

Gravitational fields always exert an **attractive force** on any object. No object is repelled when it enters Earth's gravitational field. However, magnetic and electric fields can exert **repulsive forces** as well as attractive forces.

Whether the force exerted in a magnetic or electric field is attractive or repulsive depends on the source of the field and the object placed in the field.

If a magnetic field is created by a north pole and a north pole is placed in the field, it will be repelled. A south pole placed at the same point in that field will be attracted. The easy way to remember the type of force that will occur is to use the saying 'like poles repel and unlike poles attract'. A negative charge placed in a field created by a positive charge will experience a force of attraction. Figure 8.4.2 can help you understand this.

A positive charge placed in a field created by a positive charge will experience a force of repulsion.

Each of these non-contact forces gets weaker as the distance between the source of the field and the object is increased.

Extension challenge experiment

★ ACTIVITY 8.4

Try this – tear a small corner of a single page of dry scrap paper into the smallest pieces you can manage. Ten or 20 will be sufficient. Now rub part of a plastic ruler or an inflated rubber balloon vigorously against a woollen jumper or a polar fleece. Rubbing an object like this can make it electrically charged. Hold the part of the balloon or ruler that has been rubbed about 1 cm above the pieces of paper.

Write down what you observe about the pieces of paper.

The reason the paper moves the way it does is quite complex.

In this module you have learned that:

1 like charges repel and unlike charges attract due to the electrostatic force.
2 the electrostatic force gets weaker the further apart the two charged objects are.

You also need to know that:

3 within each tiny piece of paper some (about 1 in every 100) negative charges are able to move.

These three facts are all you need to explain the observations you made about the pieces of paper.

Keep in mind that, although the balloon or ruler is charged, the pieces of paper are not charged (they are neutral).

Draw an enlarged diagram of a single piece of the torn paper. Above it, draw the ruler or balloon as appropriate to your challenge. Let's imagine that after rubbing the ruler or balloon, it has become negatively charged. Challenge yourself to use the diagram and the three facts to explain what you observed.

8.4 LEARNING CHECK

1 **State** the name of the region where a:
   a mass exerts a non-contact force on other masses.
   b charged object exerts a non-contact force on another charged object.
   c magnet exerts a force on another magnet.
2 A red balloon is given a positive charge. It is observed that a blue balloon is attracted to the red balloon but that it is repelled by a green balloon. What type of charge does the:
   a blue balloon have?
   b green balloon have?
3 **Explain** why gravity is considered to be a non-contact force. Use an example in your answer.
4 **Compare** the three types of non-contact forces to determine what magnetic and electrostatic forces have in common that is not possible for gravitational forces.
5 The needle of a compass is a small magnet that can spin on an axis. When we use a compass, it lines up so that it always points in the same direction. What can you infer about Earth from this observation?
6 **Identify** three applied non-contact force–reaction force pairs you can see or think of. Write them in the form of 'sprinter pushes back on blocks; blocks push forwards on sprinter'.

8.5 Balanced and unbalanced forces

BY THE END OF THIS MODULE, YOU WILL BE ABLE TO:
✓ explain the effect of balanced and unbalanced forces.

Quiz
Balanced or unbalanced force?

Extra science investigation
Tug-o-war forces

GET THINKING

Look at Figure 8.5.1, which shows a man pulling a tyre. What forces are acting on the tyre? Can you explain what would need to be true of those forces if the tyre is moving at a constant speed or if the tyre is slowing down? What about if the tyre is speeding up?

▲ FIGURE 8.5.1 Consider the forces on the tyre as the man drags it across the grass.

Distinguishing balanced from unbalanced forces

Figure 8.5.2 shows a street corner. All of the objects that are seen experience multiple forces acting on them. Some objects are stationary and others are moving. Whether they are stationary or moving, and how they are moving, depends on the forces acting on them.

balanced force
a force that has an equal force acting on the same object in the opposite direction

unbalanced force
a force that does not have an equal force acting on the same object in the opposite direction

In some of the situations, we would describe the forces as **balanced forces**, and in others we would describe the forces as **unbalanced forces**. You may be able to work out some of the forces acting on the objects using the knowledge you have gained so far in this chapter.

▲ FIGURE 8.5.2 A busy street corner: there are many forces acting.

Consider the forces acting on the balloon in Figure 8.5.3. The balloon would be experiencing:

- a force due to gravity acting downwards
- tension in the string acting downwards
- a force due to the air acting upwards (**buoyancy**).

The two downwards forces cancel out the upwards force and so we can say that the forces are balanced.

Consider a plastic bottle being dropped into the recycling bin, as shown in Figure 8.5.4. The force of gravity will, of course, act downwards on it. The only other force is a small force upwards caused by air resistance. Because the downwards force is greater than the upwards force, the forces do not cancel out. These forces are said to be unbalanced.

Effect of balanced and unbalanced forces

A simple way to tell if the forces acting on an object are balanced or unbalanced is to analyse the motion of the object.

- If the object is stationary, or is moving at constant speed, then the forces must be balanced.
- If the object is getting faster, slowing down or changing direction, then the forces must be unbalanced.

buoyancy
an upwards force exerted by a fluid that opposes the downwards force exerted on an immersed object

▼ FIGURE 8.5.3 A stationary balloon on the end of a string is under the influence of gravity, tension and buoyancy. The three forces are balanced.

▲ FIGURE 8.5.4 A falling bottle is under the influence of unbalanced forces.

8.5 LEARNING CHECK

1. What types of motion indicate that an object is under the influence of balanced forces?
2. **Explain** the types of motion that indicate that an object is under the influence of unbalanced forces.
3. Consider the objects observed on a street corner, illustrated in Figure 8.5.2.
    a. **Identify** two objects that are under the influence of balanced forces.
    b. **Justify** your choices in part **a**.
4. **Describe** the forces acting on the suitcase shown in Figure 8.5.2.
5. If a car is slowing down, **state** what must be true about the forces acting on the car. Consider the size of the forwards and backwards forces acting on the car.

8.6 Net force

BY THE END OF THIS MODULE, YOU WILL BE ABLE TO:
- ✓ represent forces in a diagram
- ✓ describe how to determine the net force on an object.

Interactive resource
Simulation: Find the net force

GET THINKING

Think of a time you pushed or dragged a heavy object over a rough surface. For example, you might have been pushing a box full of books across a carpet. You might have pushed at first, but it didn't move and so you pushed a bit harder, and suddenly it started moving. In terms of the forces involved, suggest why the box didn't move at first, but then moved.

Representing forces

In order to describe or analyse the forces acting on an object, it is helpful to draw a simple diagram that includes the forces. Each force is represented by an arrow that points in the direction of the force. This arrow is called a **force arrow**. The length of the force arrow indicates the strength of the force – a longer force arrow indicates a larger force. Labelling the force arrows with the type of force helps to make the diagram more informative.

force arrow
an arrow drawn on a diagram to illustrate the direction and relative strength of a force

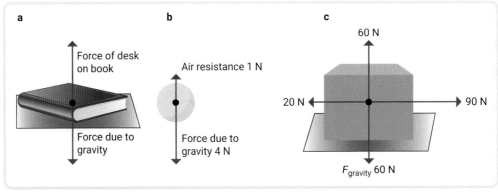

▲ **FIGURE 8.6.1** Forces acting on **(a)** a book at rest on a desk; **(b)** falling tennis ball; **(c)** box being pushed along a floor

Adding forces to find net force

total force (net force)
the sum of all forces acting on an object

Forces acting on an object add up to create a **total force** (or **net force**). If the forces cancel out, the forces are balanced and the net force is zero.

In Figure 8.6.1a, the downwards force due to gravity acting on the book is equal in size and opposite in direction to the upwards force from the desk acting on the book. The forces acting on the book are balanced and the net force on the book is zero.

In Figure 8.6.1b, the downwards force acting on the tennis ball due to gravity is much bigger than the upwards force acting on the ball due to air resistance. These forces are not balanced – there will be a net force downwards because the larger force is the downwards one. The size of the net force depends on the size of the two forces acting in this case. You can see that the upwards force due to air resistance is 1 N and the downwards force due to gravity is 4 N, so the net force is 3 N downwards.

In Figure 8.6.1c, the force upwards and the force downwards are balanced (and so the net force in the up and down direction is zero). The net force on the box can then be determined by just looking at the forwards and backwards forces. The forwards push force of 90 N is larger than the backwards frictional force of 20 N. The net force is given by subtracting the forces because they are acting in opposite directions on the box. This results in a net force of 90 N − 20 N = 70 N forwards. This is shown in Figure 8.6.2.

If two forces are acting in the same direction, then the net force can be found by adding the forces. A backwards push of 50 N and a pull in the same direction of 20 N will result in a net force of 50 N + 20 N = 70 N backwards. This is shown in Figure 8.6.3.

▲ **FIGURE 8.6.2** The forces are acting in opposite directions. Therefore, the net force is found by subtracting the backwards force from the forwards force.

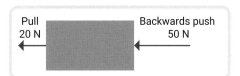

▲ **FIGURE 8.6.3** The forces are acting in the same direction. Therefore, the net force is found by adding the forces.

8.6 LEARNING CHECK

1. **Identify** another term for net force.
2. For each example, **calculate** the net force and **describe** a motion that might result from the net force.

   a

   b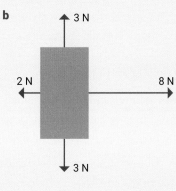

   c

3. **Draw** a labelled diagram representing the forces acting on a:
   a stationary object you can see.
   b moving object you can see.
4. **Describe** three types of motion that indicate that the net force on an object is not zero.
5. An object has the following forces acting on it: 4 N up, 6 N left, and 4 N down. What force must be added so that the net force is zero?
6. Reflect on your response to the Get thinking activity at the start of the module. Can you **explain** it better now?
7. Take a video that includes footage of a stationary object, an object moving at a constant speed and an accelerating object. Add a commentary to explain what is happening in terms of the forces acting on each object.

8.7 The effect of mass

BY THE END OF THIS MODULE, YOU WILL BE ABLE TO:
- ✓ define mass
- ✓ describe how mass affects the motion of an object experiencing a net force.

Interactive resource
Simulation: How does mass affect force?

GET THINKING

A bowling ball is smaller than a basketball and yet it is harder to pick up. What does a bowling ball have that makes this the case?

▲ FIGURE 8.7.1 Why is a bowling ball harder to pick up than a basketball?

What is mass?

mass
the amount of matter in an object, measured in kilograms (kg), grams (g) or milligrams (mg)

You will recall from Chapter 1 that **mass**, which is measured in milligrams (mg), grams (g) or kilograms (kg), is the amount of matter that makes up an object. Objects that have more matter have greater mass. As is shown in Figure 8.7.2, an apple has more matter, and therefore mass, than a cupcake, even though they may be the same or similar size.

The effect of mass

You know from previous modules that when an object experiences an unbalanced or net force, its motion changes – it speeds up, slows down or changes direction. How much it changes will depend on the mass of the object.

Imagine a toy car and a real car rolling down a hill towards you. It will be much easier to apply a force to stop the toy car than to stop the real car. This is because the toy car as shown in Figure 8.7.3 has a much smaller mass.

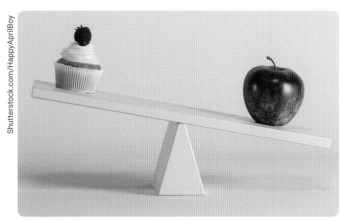

▲ FIGURE 8.7.2 The apple has more mass than the cupcake even though they are a similar size.

▲ FIGURE 8.7.3 Less force would be needed to stop the toy car than to stop the heavier real car.

We can express this in a general statement: if the same net force is applied to two objects of different mass, the one with the smaller mass will experience a greater change in motion.

This explains why it is easier to throw a tennis ball than a cricket ball, why it would hurt less if a peanut fell on you than if a coconut did, why it is easier to push a child on a swing than a large adult, and why it easier to pick up a pebble than a boulder.

▲ FIGURE 8.7.4   It is easier for the woman to change the motion of the child on the swing by applying a force than it would be for her to change the motion of an adult on the swing by applying a force. This is because the child has less mass.

8.7 LEARNING CHECK

1. **Write** a sentence to tell someone what mass is, without using the word 'matter'.
2. Which has more mass – a kilogram of feathers or a kilogram of metal? **Justify** your answer.
3. Does air have mass? **Discuss** with a friend. **Justify** your response.
4. Truck drivers know to leave a bigger gap between them and the vehicle in front when they have a full load than when their truck is empty. **Explain** why this is important for safety.
5. **Explain** why it is easier to swing a bucket in a circle when it is empty than when it is full of water.
6. Extension: Consider a heavy object that you can move by hand. This might be a 2 L carton of milk, a bowling ball, a big book or something completely different but similarly heavy.

   Now imagine lifting the object, with one hand, from the floor to above your head. Think about how much effort that would require.

   Next imagine pushing the object, with one hand, to make it slide away from you as fast as you could across a very, very smooth surface. Think about how much effort that would require.

   If you repeated each of these two 'thought experiments' inside a spaceship in deep space – a long way from any stars or planets – would each require the same effort, more effort or less effort? Why? Try to use the word 'mass' in your response.

8.8 Mass and weight

BY THE END OF THIS MODULE, YOU WILL BE ABLE TO:
- ✓ calculate the weight of an object of known mass in a given gravitational field
- ✓ explain why mass remains constant but weight may change when location changes.

Video activity
Measuring weight in space

GET THINKING

Figure 8.8.1 shows an astronaut walking on the Moon. How would walking on the Moon differ from walking on Earth? Why?

▶ **FIGURE 8.8.1** Walking on the Moon is different from walking on Earth.

Weight versus mass

A 1.5 kg tub of ice-cream would have a mass of 1.5 kg here on Earth and on the Moon. The mass of the ice-cream stays the same because the amount of matter is the same. However, the **weight** of the ice-cream on the Moon would be much less than it is on Earth.

weight
the downwards force on an object due to gravity

Weight is the downwards force on an object due to gravity. We also describe weight as the force experienced by an object in a gravitational field. Weight is a force and, therefore, is measured in newtons. All objects around you have a:
- mass because they have matter
- weight due to the force exerted by gravity.

Changing weight

Because weight is the force downwards on an object due to gravity, it depends on the mass of the object and the amount of gravity. The amount of gravity is described by the term '**acceleration due to gravity**'. This is a measure of how quickly the speed increases when an object falls. The acceleration due to gravity, represented by the symbol 'g', is described by a number and the units $m\,s^{-2}$ (metres per second squared). The acceleration due to gravity is different in in different places in the solar system, although it is nearly the same at all places on Earth.

acceleration due to gravity
the rate at which a falling object gets faster due to the force of gravity

Weight can be calculated by multiplying mass by acceleration due to gravity.

Weight = mass × acceleration due to gravity

The acceleration due to gravity on Earth is $9.8\,m\,s^{-2}$. So, the weight of a 5 kg mass on Earth can be calculated using the equation above.

Weight = $5\,kg \times 9.8\,m\,s^{-2}$ = $49\,N$

The acceleration due to gravity on the Moon is 1.6 m s⁻². So, the weight of a 5 kg mass on the Moon is:

Weight = 5 kg × 1.6 m s⁻² = 8 N

Extending this idea, we can see that we can move an object that has a constant mass to a range of places with different gravity (such as the Moon, Mars, Mercury and Jupiter) and the weight will be different in each place.

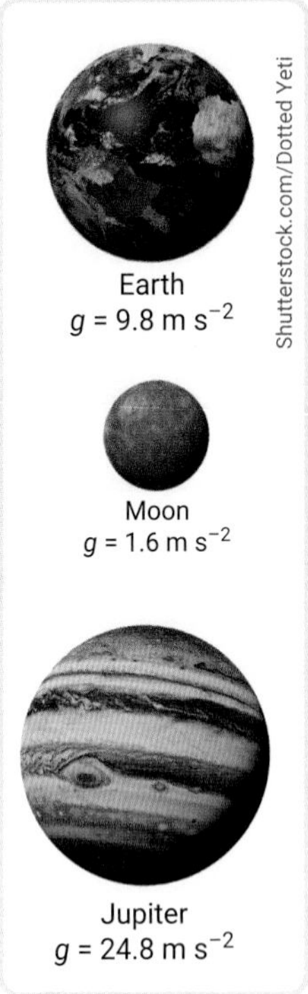

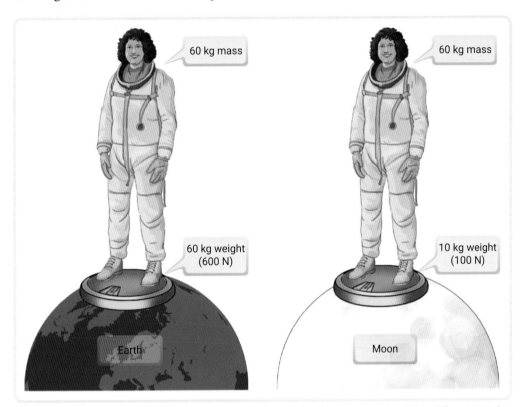

▲ FIGURE 8.8.2 Gravity has different values on Earth, the Moon and Jupiter.

Earth
$g = 9.8$ m s⁻²

Moon
$g = 1.6$ m s⁻²

Jupiter
$g = 24.8$ m s⁻²

▲ FIGURE 8.8.3 An astronaut will have the same mass on the Moon as on Earth but a different weight.

Weightless or zero gravity

Perhaps you have watched a science fiction movie that has scenes in which the characters or objects in or around a spacecraft seem to float as if they are weightless or there is no gravity.

Movies such as *Gravity*, *The Martian* or *Apollo 13* all feature very realistic and scientifically accurate versions of this.

Alternatively, YouTube videos show astronauts on the International Space Station experiencing this apparent weightless or zero-gravity feeling.

What to do

The terms 'weightlessness' and 'zero-gravity' are often used to describe these scenes, but are these terms correct? Using what you have learned in this module about weight and gravity, discuss with a classmate whether having no weight or finding a place with no gravity is possible.

☆ ACTIVITY

▲ FIGURE 8.8.4 Astronauts on the International Space Station experience apparent weightlessness.

In your discussion, you might have considered the Moon, which orbits Earth because of the effect of Earth's gravity, or the planets of our solar system, which orbit the Sun because of the effect of the Sun's gravity. This tells us that there is gravity acting throughout our solar system. The International Space Station (ISS) is only a few hundred kilometres from Earth's surface (the movie *Gravity* is set in a similar location).

How does it seem that there is no gravity and, therefore, no weight on the ISS and in the movie *Gravity*?

To help you answer that question, imagine a situation in which you are in an elevator at the top of the tallest building in the world. You are standing on a set of bathroom scales, and you are holding an apple in your hand in front of your face. You can feel the weight of the apple and you know if you let go of it the apple would fall to the floor of the elevator. The bathroom scales indicate your mass because you are pressing down on them – you can feel that pressure in your feet.

Suddenly the cable holding the elevator breaks and you are falling straight down. As you fall, you notice that the 'feel' of the apple has changed, as has the feel of your feet on the scales. If you look down at the scales, what would you see? If you dropped the apple, what would you see?

What do you think?

Discuss this with a classmate.

Check with the teacher to see if you are correct.

Now look at the question about the ISS again. How would you explain the apparent weightlessness to a friend or family member now?

8.8 LEARNING CHECK

1. Write two different phrases to describe 'acceleration due to gravity' in your own words.
2. **Discuss** whether we use the term 'weight' correctly.
3. **Calculate** the weight of a 50 kg boy on Mars where the acceleration due to gravity is 3.7 m s^{-2}.
4. **Compare** the meaning of the terms 'weight' and 'mass'.
5. Use what you learned in Module 8.4 about how Earth's gravitational field changes as you get further from Earth to **predict** where on Earth your weight would be least. **Explain** your answer.
6. Challenge: **Calculate** the mass of an object if its weight is 120 N on Earth.

8.9 Simple machines

BY THE END OF THIS MODULE, YOU WILL BE ABLE TO:
- ✓ identify simple machines
- ✓ explain how simple machines can change the size of a force needed.

GET THINKING

Skim read the module, paying particular attention to the headings and images. Create a table or mind map with the name and description of each type of simple machine. Add extra information to the table or mind map as you learn about them during the module.

Video activity
Levers, wheels and pulleys

Interactive resource
Match: Simple machines

What is a simple machine?

Have you ever wondered how a wheelbarrow or scissors make life easier for us?

Throughout history, people have developed machines to make difficult tasks easier, or otherwise impossible tasks achievable. Hammers, chisels, scissors, ramps and wheelbarrows use **simple machines** to perform their function. There are six simple machines from which machines are designed: **inclined plane, wedge, screw, wheel and axle, pulley** and **lever**.

The main purpose of a simple machine is to increase the size of the applied force. For example, a 10 N applied force can move an object that would otherwise require 50 N of force to be moved. Simple machines can also perform one or more of the following functions at the same time:
- enable a force to be applied in one place but act in another place
- change the direction of a force.

How do simple machines help us?

The amount by which a simple machine increases the size of the applied force is known as the machine's **mechanical advantage**.

If we wanted to lift a heavy object into the back of a truck, we would need to apply a force (to overcome the force of gravity) over the distance from the ground to the back of the truck. When a force is applied to an object as it moves over a distance, **work** is done. The amount of work done is the product of the force applied and the distance moved. Work is measured in units of joules (J).

$$\text{Work} = \text{force} \times \text{distance}$$

For example, if a 20 N force is applied to a box that slides 5 m, then the work done is given by:

$$20\,\text{N} \times 5\,\text{m} = 100\,\text{J}$$

It is important to note that the same amount of work needs to be done to lift the heavy object into the truck no matter the path that we take.

For example, we could move the heavy object up a smooth ramp instead of lifting it straight up. Since the amount of work remains the same, if the distance travelled is doubled, the force needed is halved. Therefore, it is much easier to get the heavy object into the truck.

simple machine
a device to increase the size of an applied force

inclined plane
a sloping ramp

wedge
a triangular-shaped tool tapering to a thin edge that acts as a portable inclined plane

screw
a long, inclined plane wrapped around a solid cylinder

wheel and axle
a solid rod connected to a wheel

pulley
a wheel on an axle that enables a change in direction of a rope or cable

lever
a solid plank or bar that rotates about a point

mechanical advantage
a measure of the force multiplication provided by a machine

work
the energy transferred to or from an object when a force is continuously applied to the object over a distance as the object moves

In this case, the mechanical advantage is 2 because the force needed to complete the job without a simple machine is twice the force needed with a simple machine.

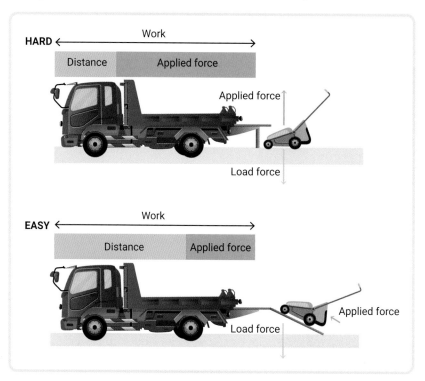

▲ FIGURE 8.9.1 A smaller applied force over a greater distance does the same amount of work as a greater applied force over a shorter distance.

Types of simple machines

Inclined plane

An inclined plane is a flat surface at an angle; for example, a ramp used for wheelchair access to a building.

The amount of work required to move an object is unchanged, but an inclined plane enables a smaller force to be applied to move an object vertically even though it is over a longer distance. The longer the inclined plane, the less steep it is and the greater the mechanical advantage. Therefore, less force will be required to move an object up it but the distance the object will need to be moved will increase.

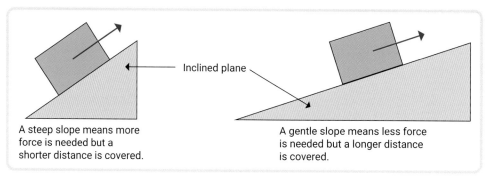

▲ FIGURE 8.9.2 An inclined plane gives the greatest mechanical advantage if it is long and less steep.

Wedge

A wedge is a solid triangular object tapering towards one edge. A wedge is like a moving inclined plane that can be driven between two objects to exert a force that separates them. For example, the blade of an axe is a wedge.

The amount of work required is unchanged – a wedge might need to be moved a long way to achieve a small lift, but the force required will be less than needed to lift the object without a wedge. The further the wedge moves to achieve a lift, the greater the mechanical advantage.

A wedge also changes the direction of the force. When a wedge is pushed forwards between a heavy object and the floor, the object is forced upwards. When a wedge is pushed downwards into a log (as seen in Figure 8.9.3), the pieces of wood are pushed sideways away from each other. The narrower the wedge, the greater the mechanical advantage.

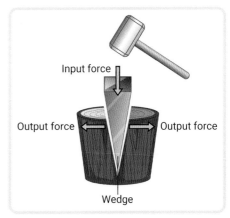

▲ **FIGURE 8.9.3** A wedge such as this multiplies the force and changes the direction of the force. Here, the sledgehammer pushes the wedge downwards, and the wedge pushes the wood outwards, causing the wood to split.

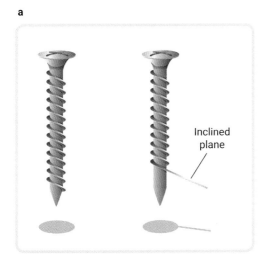

▲ **FIGURE 8.9.4** (a) A screw consists of an inclined plane wrapped around a cylindrical shaft. (b) A mountain road is like a screw where an inclined plane changes direction.

Screw

A screw is a long, inclined plane wrapped around a solid cylinder. Therefore, the way a screw works is similar to the inclined plane. For example, a car jack is a screw used as a simple machine.

The amount of work required is unchanged – each turn of the screw moves it a small distance. However, as the force is applied over a larger distance as the screw is turned, the force needed is less. This is evidence of the mechanical advantage of the screw. A winding road up a mountain has features in common with a screw.

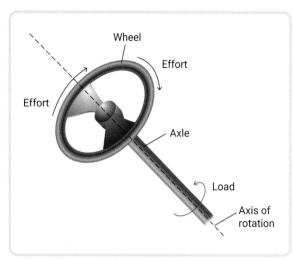

▲ FIGURE 8.9.5 A wheel and axle multiplies force and changes the position of the force.

Wheel and axle

If a large wheel is connected to a small axle, when the wheel is turned, the axle will also turn. For example, a bathroom tap is a wheel and axle.

The amount of work required is unchanged – the turning force that the axle can apply is greater than the turning force needed to turn the wheel. This is the mechanical advantage of the wheel and axle system. The bigger the radius of the wheel compared with the axle, the greater the mechanical advantage. Trucks and buses have very large steering wheels to use the mechanical advantage offered by this simple machine. As with other simple machines, a smaller force needs to be applied over a greater distance. The outside of the wheel will move a larger distance in a circular arc than the small distance moved by the outside of the axle.

Pulley

A pulley is an arrangement of a wheel that can turn on an axis and is used to change the direction of a tension force in a rope or string. For example, a pulley is used when a bucket is raised from a well.

The more pulleys that are used in combination, the greater the mechanical advantage. This means that a smaller applied force can result in the required force although the amount of work required is unchanged. The distance the rope must be pulled will increase by the same value as the mechanical advantage. For example, an object requires a force of 150 N to be moved and needs to be moved 2 metres. The mechanical advantage of the pulley system is 3. Only 50 N must be applied to move the object, but the 50 N force must be applied over 6 metres. The calculation is as follows.

▲ FIGURE 8.9.6 A pulley multiplies force and changes the direction of the force.

Work = force × distance
  = 150 N × 2 m
  = 300 J

Mechanical advantage = 3

Force required is reduced to $\frac{150}{3} = 50$ N

Same amount of work, therefore work = 300 J

300 J = 50 N × distance

Therefore, distance = $\frac{300}{50} = 6$ m

Lever

A lever is a solid straight object, like a piece of wood or a metal bar, that rotates about a point called a **fulcrum**. For example, a seesaw is a lever.

fulcrum
the point at which a lever is supported and rotates

A small force applied at a large distance from the fulcrum results in a large force acting at a small distance from the fulcrum. The bigger the distance from the applied force (or effort force) to the fulcrum compared with the distance from the load to the fulcrum, the greater the mechanical advantage. A seesaw, a spoon being used to remove a tight lid from a food tin, and a screwdriver being used to remove the lid of a paint can are examples of levers.

The amount of work required is unchanged, but a smaller force applied, over a longer distance, results in a larger force applied over a smaller distance.

For example, the lever in Figure 8.9.7 has a mechanical advantage of 4. So, if a 100 N force is required to lift the box directly, then the applied force on the lever needs only to be 25 N to achieve the same result. However, if the box needs to be lifted 30 cm then the applied force would need to be applied over a distance of 4 × 30 cm (120 cm) to lift the box the required distance.

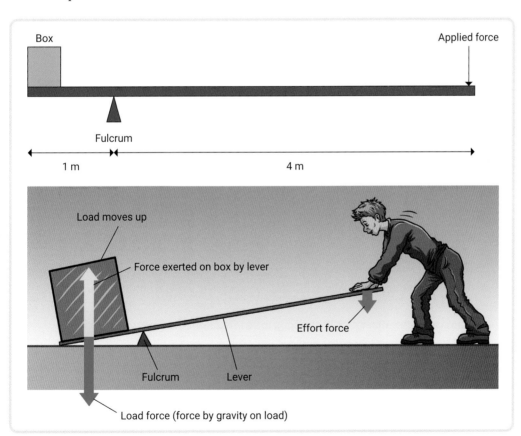

▲ **FIGURE 8.9.7** A lever multiplies force and changes the position and direction of the force.

8.9 LEARNING CHECK

1. **Describe** the main function of a simple machine.
2. If you have tried to ride a bicycle up a steep hill, you might have ridden from side to side in a 'zigzagging' way to make it easier. **Compare** this process with using a screw, a ramp, a wheelbarrow or a pair of pliers.
3. **State** what is meant by mechanical advantage.
4. Copy and complete the following table by placing an 'x' in the appropriate column to identify what type of simple machine each device is.

Machine/device	Inclined plane	Wedge	Screw	Wheel and axle	Pulley	Lever
Axe						
Bottle opener						
Car jack						
Cheese grater						
Crowbar						
Door hinge						
Doorknob						
Drill						
Flag raising device on a flagpole						
Jar lid						
Ladder						
Pliers						
Screwdriver						
Shovel						
Tweezers						
Wheelbarrow						

5. Research Archimedes' screw and write a paragraph to **explain** how it works and what it is used for.
6. A boulder with a mass of 120 kg needs to be lifted from the ground.
    a. **Calculate** the weight of the boulder.
    b. **Calculate** the force needed to lift the boulder with a lever that has a mechanical advantage of 15.
    c. Challenge: If the end of the lever is pushed down 1208.9cm, how far will the boulder be lifted? (Hint: Calculate the work done to push the lever down and use this to calculate the distance the boulder moved.)

FIRST NATIONS SCIENCE CONTEXTS

8.10 First Nations Australians' spear-throwing technology

IN THIS MODULE YOU WILL: ✓ investigate the use of forces in spear-throwers developed by First Nations Australians.

Examining spear-throwers

First Nations Australians developed and produced a range of tools designed to increase both the speed and accuracy of hunting projectiles, such as spears and arrows. Across much of Australia, spear-throwers were developed and refined by many First Nations Australian Peoples. Bows and arrows were used as well as spear-throwers by Peoples of the Torres Strait.

Records indicate First Nations Australians began to manufacture and use spear-throwers at least 5000 years ago and that the development of this hunting technology is similar to that of other First Nations Peoples of the world.

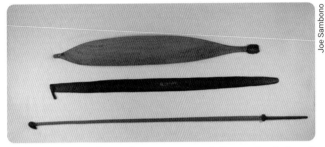

▲ **FIGURE 8.10.1** Spear-throwers are carefully constructed, and different cultural groups use different designs. Origin from top to bottom: Central Desert Region, Northern Territory; Cape York, Queensland; Port Keats, Northern Territory.

Spear-thrower technology highlights First Nations Australians' understanding of forces and levers. A spear-thrower helps hunters throw further by acting as an extension of the thrower's arm. A longer 'arm' (or longer lever) means that more force can be put behind the spear and it can go faster and further.

Spear-throwers are crafted out of wood, usually hardwood such as mulga. However, the shape, width and length vary depending on the cultural group's design (see Figure 8.10.1) and the environment in which it was used. A spear-thrower needs to be carefully constructed for each individual user to maximise its effect when used.

Spear-throwers typically narrow towards the gripping end where notches coated in resin are cut into either side to form a hand grip (see Figure 8.10.2a). The other end has a 'peg', often made from a different material, that fits into a socket or notch at the base of the spear (see Figure 8.10.2b).

▶ **FIGURE 8.10.2** (a) Spear-throwers are used to increase the distanced travelled, and the speed and accuracy of spears. This spear-thrower is from Galiwinku, Northern Territory. (b) The peg at the base of the spear.

ACTIVITY

1 **Draw** a labelled force diagram to show the forces acting on a spear:
   a when it is resting in the spear-thrower (before being thrown).
   b in motion through the air after being thrown.
2 For the diagrams you drew in Question 1, **explain** if the forces are balanced or unbalanced.
3 **Describe** what advantages the spear-throwers give to the hunters.
4 **Explain** how the spear-throwers give these advantages. (Hint: Spear-throwers are levers.)

8.11 Principles of rocket propulsion

SCIENCE AS A HUMAN ENDEAVOUR

BY THE END OF THIS MODULE, YOU WILL BE ABLE TO:
- ✓ explain the forces involved in rocket propulsion

Video activity
Rockets 101

▲ **FIGURE 8.11.1** Hero's aeolipile used the principles of rocket propulsion.

Rockets uses simple principles

The rockets that are used today to launch people, equipment, space probes and satellites from Earth's surface into space are remarkable examples of the development of ideas over time. Despite their complexity and impressive capability, rockets use simple principles to achieve extraordinary outcomes.

History of rocket development

The early history of the development of the rocket is not completely clear; however, one of the key principles associated with rocket propulsion was first demonstrated around 100 BCE by a Greek inventor, Hero of Alexandria. His device was called an aeolipile and it used steam as its method of propulsion. You can see a representation of Hero's device in Figure 8.11.1.

Hero mounted a metal sphere on pipes above a large tub of boiling water heated by a fire. The sphere was able to rotate around the end of the pipes. The boiling water created steam, which flowed along the pipes and into the sphere. The sphere had two L-shaped pipes on opposite sides from which the steam could escape. His demonstration amazed onlookers at the time.

8.11 LEARNING CHECK

1. When the steam is coming from the two L-shaped pipes, as seen in Figure 8.11.1, **describe** what would happen to the metal sphere.
2. **Describe** the applied force and reaction force occurring in the aeolipile.
3. **Explain** how this relates to the propulsion of rockets. Use a diagram and refer to action force and reaction force in your response.

SCIENCE INVESTIGATIONS

8.12 Analysing data in tables and graphs

SCIENCE SKILLS IN FOCUS

IN THIS MODULE, YOU WILL FOCUS ON LEARNING AND IMPROVING THESE SKILLS:

- organise data in an effective table
- analyse data using a graph to determine a mathematical relationship.

Tables and graphs are important and powerful tools for organising and analysing data. In Chapters 2 and 3 you looked at making tables and graphs. In this module, you will practise and refine those skills.

Recall that all tables have:
1. an informative title
2. ruled columns
3. column headings that describe the data
4. units for the data in the column in brackets next to the column heading only
5. the independent variable (the one you changed) in the first column
6. the dependent variable (the one you measured) in the next columns.

Recall that all graphs have:
1. an informative title
2. ruled axes
3. a numerical scale that increases in even increments
4. labels for axes that describe the data
5. units for axes in brackets next to the label
6. the independent variable (the one you changed) on the horizontal axis
7. the dependent variable (the one you measured) on the vertical axis
8. average results only (in most cases)
9. a line (or curve) of best fit.

Always make your graph occupy as much of the grid space as possible.

THE RELATIONSHIP BETWEEN WEIGHT AND FRICTION

AIM

To determine the relationship between the weight of an object and the force of friction that acts on the object

YOU NEED

- ☑ spring balance
- ☑ wooden block with a hook on one end or string attached to it
- ☑ frictional surface such as a wooden plank, piece of carpet or piece of sandpaper
- ☑ range of masses increasing in small increments from 250 g to 1 kg
- ☑ electronic balance

WHAT TO DO

1. Set up the equipment as shown in Figure 8.12.1.
2. Draw a table to record the force needed to move the block with different added masses.
   a. Construct a table with the total mass in the first column. Place the calculated weight in the second column and the force measured in the third column. Divide the third column into four sub-columns for trial 1, trial 2, trial 3 and average values. Include the title and unit for each column.
   b. Ensure you have enough rows for the number of masses you will be able to add.
3. Measure and record the mass of the block on the electronic balance.

Video
Science skills in a minute: Representing data

Science skills resource
Science skills in practice: Collecting and representing data

Chapter 8 | Forces

8.12

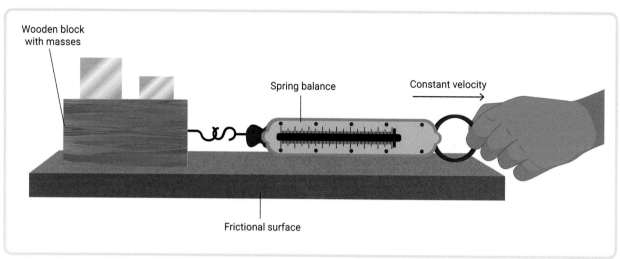

▲ FIGURE 8.12.1 A spring balance is used to measure the force required to pull a mass along a surface.

4 Attach the spring balance to the hook or string on the block and slowly and steadily drag the block along the frictional surface. Record the force required to move the block. Note that this force may vary, up and down, during the motion. Record the value that represents the middle or most consistent value.

5 Add a mass to the block.

6 Measure and record the mass of the block plus one added mass on the electronic balance.

7 Repeat step 4.

8 Continue to add masses and record force values until you have used all the masses or have at least five separate mass values.

9 Repeat the entire process so that you have three trials for each of the mass values you use.

10 Calculate the weight of each of the mass values in the first column. Record these values in the second column.

WHAT DO YOU THINK?

1 Why does the value of the force recorded by the spring balance vary a little as you drag the mass?

2 What was the purpose of repeating the experiment three times for each mass?

3 Calculate the average value of the force recorded for each mass and complete that column of the table.

4 Draw a graph of the weight being pulled and the average force required.
    a Label and include units for the axes and give the axes an even scale so that your data to take up as much of the graph as possible.
    b Include an appropriate title (which can be the same as the table).
    c Plot your data.
    d Draw a line of best fit.

5 What relationship does the graph suggest between the weight of the pulled object and the average force needed to pull it at a steady rate?

6 How did plotting a graph and drawing a line of best fit help identify the trend?

7 How can you explain your results in terms of the model of friction described earlier in this chapter?

8 REVIEW

REMEMBERING

1 **List** the three non-contact forces.
2 **State** the general name given to forces that act along a string or rope or cable.
3 **State** the name of the force that is acting on the Moon as it orbits Earth.
4 **Define** 'net force'.
5 **Identify** the statements below as true or false.
   a Stationary objects have no forces acting on them.
   b A force is an interaction between objects.
   c Both mass and weight would change if an object was moved from Earth to the Moon.
   d Force is measured in kilograms.
   e There are always two forces acting on any object – the applied force and the reaction force.
   f The more mass an object has, the harder it is to change its motion

UNDERSTANDING

6 **Justify** why skiers and speedskaters in the Winter Olympics wear shiny, smooth suits and helmets and tuck down as they move.

7 **Describe** the effects of an unbalanced force.
8 **Describe** how you presume oil reduces the friction between two surfaces.
9 **Identify** a situation in which an object could have no weight.
10 **Explain** how a spring balance measures force.
11 **Explain** how the mass of a spear affects the distance it travels when a spear-thrower is used.

APPLYING

12 At high altitudes, such as near mountain tops, there is less air. When long-jump and cycling competitions take place at high altitudes, world records are often set. **Explain** why this is the case, using the ideas you have learned about friction.
13 **Predict** what would happen if you tried to jump on the Moon.
14 **Compare** the force exerted by a magnetic north pole on another north pole with the force exerted by the Sun on Earth.
15 **Describe** a situation in which you have relied on friction to complete a task today.
16 **Predict** what will happen if you increase the length of a lever while keeping everything else the same.
17 **Explain** how you could increase friction between your feet and the ground when walking.
18 **Classify** the following situations as having balanced forces or unbalanced forces.
   a A kite motionless in the air
   b A leaf falling at a steady rate
   c A ball going over the fence and landing in the neighbour's pool
   d The ball floating in the neighbour's pool
   e An elevator moving up ten floors going past one floor of a building every 3 seconds
   f A cyclist going around a corner
   g A snowboarder slowing down at the end of a run
   h A weights bar held above a weightlifter's head

Chapter 8 | Forces

19 You need to move a heavy box of books across the room. Without using any simple machines, **describe** three changes you could make so that the box will cross the room in less time.

20 **Describe** a situation around the house or school in which a pulley is used.

21 A tennis ball falls towards the ground. **Draw** a diagram with force arrows to show the contact force and non-contact force that act on the ball, clearly indicating which force is larger.

EVALUATING

22 **Explain** why it is easier to move an object up a set height by using a long ramp than by using a short ramp.

23 **Predict** the changes to the weight of a lunar rover when it is transported from the surface of Earth to the surface of the Moon.

24 **Apply** your knowledge of forces and simple machines to explain why less force is needed to turn a screwdriver with a wide handle than one with a narrow handle.

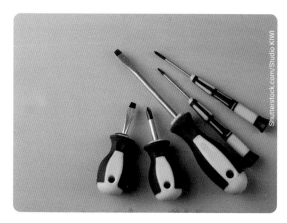

25 **Predict** what difference you would notice when driving two identical screws into the same piece of wood if one has five turns per centimetre and the other has 10 turns per centimetre. **Justify** your prediction.

26 **Evaluate** the statement attributed to Archimedes: 'Give me a lever long enough and a fulcrum on which to place it and I shall move the world'.

27 Consider the simple machines described in this chapter. Which do you think has affected our lives the most? **Justify** your choice.

CREATING

28 **Sketch** a horse pulling a sled along the ground. Annotate the diagram with all the contact and non-contact forces acting on the horse, the sled and the ground.

BIG SCIENCE CHALLENGE PROJECT #8

1 Connect what you've learned

In this chapter, you have learned about the forces and how they do or do not change the way an object moves.

Draw a mind map to show the types of forces and the circumstances when forces cause motion to change, or not change. Use the keywords from the start of each module to prompt you.

2 Check your thinking

At the start of the chapter, you were asked the following questions.

- How could an object so huge be moved by people without sophisticated machinery?
- How can we use a knowledge of forces to overcome the challenge to lift and move heavy objects over a distance?

Use the knowledge gained in this chapter to explain the forces that need to be overcome to lift and move heavy objects. Describe how this can be done with simple machines.

3 Make an action plan

Research how scientists believe the Easter Island stones, and other ancient heavy statues and objects (including Stonehenge, Angkhor Wat and the pyramids), were moved by people without sophisticated machinery. Where can you see the principles that you have learned about forces and simple machines represented in your research?

4 Communicate

Make a poster or presentation that could be used to teach a Year 5 or 6 class about forces.

9 Our place in space

9.1 Earth, the Sun and the Moon (p. 256)
Earth orbits the Sun, and the Moon orbits Earth.

9.2 Reviewing the rotation of Earth (p. 258)
Earth rotates on its axis once every day.

9.3 Earth's revolution around the Sun (p. 260)
Earth revolves around the Sun once every year.

9.4 Seasons (p. 263)
Seasons are caused by the tilt of Earth's axis as it moves around the Sun.

9.5 Phases of the Moon (p. 266)
Phases of the Moon occur when we see different amounts of the lit face of the Moon.

9.6 Eclipses (p. 269)
Eclipses are caused when light from the Sun is blocked by another object.

9.7 Tides (p. 273)
Tides are the result of gravitational attraction on the ocean by the Moon and Sun.

9.8 FIRST NATIONS SCIENCE CONTEXTS: First Nations Australians' knowledge of Moon phases and tides (p. 276)
First Nations Australians have observed the night sky and built a wealth of astronomical knowledge over thousands of years.

9.9 SCIENCE AS A HUMAN ENDEAVOUR: Galileo's testing of scientific ideas (p. 279)
Galileo was sent to prison for stating that Earth orbited the Sun.

9.10 SCIENCE INVESTIGATIONS: Modelling data (p. 280)
1 Modelling the behaviour of Jupiter's moons: a mathematical model
2 Earth's tilt and illumination: a physical model
3 Modelling phases of the moon: a physical model

BIG SCIENCE CHALLENGE

▲ FIGURE 9.0.1   Sunrise

Each day, as the Sun rises, it results in changes on Earth's surface.

▸ Would days and nights on Earth be different if Earth was a little further away from the Sun?

▸ How would the oceans be different if Earth had two moons, like Mars?

▸ What if the Moon was further away from Earth than it is now? What would change on Earth?

Are you ready to learn how to answer these questions?

#9 SCIENCE CHALLENGE ACCEPTED!

At the end of this chapter, you will complete the Big Science Challenge Project #9. You can use the information you learn in this chapter to complete the project.

Assessments
- Prior knowledge quiz
- Chapter review questions
- End-of-chapter test
- Portfolio assessment task: Data test

Videos
- Science skills in a minute: Modelling data **(9.10)**
- Video activities: What is an orbit? **(9.1)**; Why does Earth have seasons? **(9.4)**; What are eclipses? **(9.6)**; The Moon and spring tides **(9.7)**

Science skills resources
- Science skills in practice: Modelling data **(9.10)**
- Extra science investigations: Modelling eclipses **(9.6)**; The effect of the Moon on tides **(9.7)**

Interactive resources
- Match: Phases of the Moon **(9.5)**
- Crossword: Moving Earth **(9.3)**

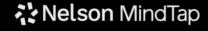

To access these resources and many more, visit
cengage.com.au/nelsonmindtap

9.1 Earth, the Sun and the Moon

BY THE END OF THIS MODULE, YOU WILL BE ABLE TO:
- ✓ identify the orbits of Earth and the Moon
- ✓ describe the position of Earth, the Sun and the Moon relative to each other.

Video activity
What is an orbit?

GET THINKING

How do the Sun, Earth and the Moon move relative to each other? Draw a labelled diagram to show your answer. How confident are you in your answer? Explain why you are, or are not, confident.

Planets and moons

planet
a natural body that orbits a star

orbit
the regular, repeating path an object takes in space around another object

satellite
an object in orbit around a larger object

moon
a natural satellite of a planet

How are planets and moons different from each other? A **planet** is a natural body that travels around a star in a regular, repeating path called an **orbit**. Earth is a planet because it orbits the star we call the Sun. Other planets that also orbit the Sun are Mercury, Venus, Mars, Jupiter, Saturn, Uranus and Neptune.

Any object in orbit around a larger object is called a **satellite**. Therefore, Earth is a satellite of the Sun. If the object is a natural satellite of a planet, it is called a **moon**. There are more than 200 moons in our solar system. Earth only has one moon, Jupiter has 79 moons, but Mercury and Venus do not have any moons. Because Earth only has one moon, we refer to it as 'the Moon' (see Figure 9.1.1).

In summary, a planet is a satellite that orbits the Sun and moons are satellites that orbit planets.

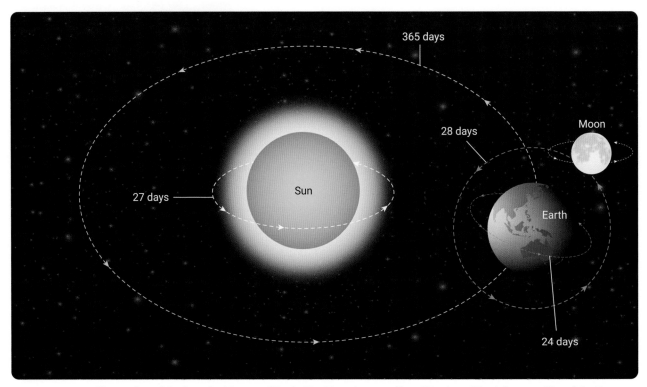

▲ **FIGURE 9.1.1** The motion of the Sun, the Moon and Earth. The Moon is a satellite of Earth and Earth is a satellite of the Sun.

▲ FIGURE 9.1.2 The Moon. As the Moon orbits Earth, it keeps the same face towards Earth (you will learn more about this in Module 9.5).

▲ FIGURE 9.1.3 Earth and the Moon are a long way from each other.

Distances in space

The distances involved in orbits are extremely large. The average distance of the Moon to Earth is 384 400 km. The average distance of Earth to the Sun is 149 600 000 km – 389 times greater than the distance of the Moon to Earth. It is very hard for us to understand such large distances. Figure 9.1.3 shows the Moon and Earth together; however, the image was taken about 541 000 km from Earth!

9.1 LEARNING CHECK

1. **Define** 'orbit'.
2. **Describe** the orbits of Earth and the Moon.
3. **Draw** a labelled diagram to show the orbits of Earth and the Moon.
4. **Create** an animation or model to show the orbits of Earth and the Moon.
5. **Explore** the moons of other planets. **Summarise** the number of moons for each planet in a table.
6. **Explain** why it may be difficult to create a diagram that accurately shows the distances between the Sun, planets and the Moon.

9.2 Reviewing the rotation of Earth

BY THE END OF THIS MODULE, YOU WILL BE ABLE TO:
- ✓ describe the rotation of Earth on its axis
- ✓ explain the cause of daytime and night-time.

GET THINKING

How are day and night related to the movement of the Sun and stars? Scan the headings and images in this module and note down what you think the module is about.

Why we have daytime and night-time

Quiz
Earth rotation

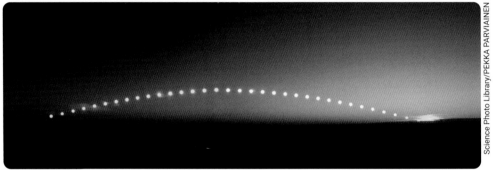

▲ FIGURE 9.2.1 The movement of the Sun across the sky during a day. The height of the path varies during the year.

Planets have an imaginary line through their centre that they spin, or rotate, around. This is called the **axis**, or rotational axis. Earth's axis has a **tilt** of 23.5° (see Figure 9.2.2).

We have **daytime** and **night-time** because Earth rotates about its axis. As Earth turns, a place in darkness moves into the light of the Sun. In this lit area, it is daytime. In the shadow on the other side of Earth, it is night-time. Earth spins from west to east, which is why the Sun appears to rise in the east and set in the west.

axis
an imaginary line that an object spins around

tilt
sloping away from the vertical

daytime
the time of day between sunrise and sunset

night-time
the time of day between sunset and sunrise

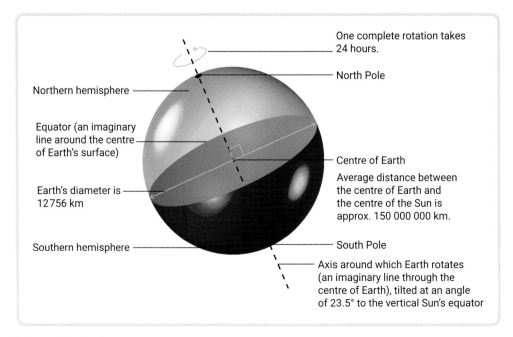

▶ FIGURE 9.2.2 Earth's geometry

258 Nelson Science 7 | Australian Curriculum 9780170472852

A day

When a planet or moon spins once on its axis, we say it has made a **rotation**. The time for Earth to make a single rotation on its axis (24 hours) is called a **day**. Other planets spin at different speeds, so their day lengths are different. For example, Jupiter, although many times larger than Earth, spins faster and only takes 9 hours and 56 minutes to rotate on its axis.

rotation
the motion of an object around an internal axis

day
the time it takes for a planet to make one rotation around its axis

The rotating Earth

Earth rotates on its axis once every 24 hours. We do not notice Earth spinning because everything around us, including the atmosphere, is spinning with Earth. The surface of Earth at the equator is travelling at a speed of 1600 km h^{-1}! The speed decreases as you move to the poles because the distance travelled becomes shorter. To the south, in Melbourne, the surface spins at only 1317 km h^{-1}.

Earth's tilt and day length

The length of daytime and night-time depends on where you are on Earth and the time of the year. Near the equator, the length of daytime is about 12 hours and 7 minutes and varies by only 2 minutes throughout the year. As you move south or north, the length of daytime and night-time changes. In the Arctic and Antarctica, daytime and night-time can be as long as 24 hours. During winter, daytimes are shorter than they are in summer. Night-times are longer in winter and shorter in summer. You will learn more about the cause of this later in the chapter when we examine the seasons.

◀ **FIGURE 9.2.3** An Antarctic sunset. In mid-winter, the Sun does not rise for a couple of weeks.

9.2 LEARNING CHECK

1. **State** the angle of tilt of Earth's axis.
2. **State** how long it takes Earth to rotate once on its axis.
3. How is daytime different from a day?
4. How does Earth's rotation cause daytime and night-time?
5. **State** two factors that affect the length of day and night.
6. Research the longest and shortest day length where you live. **Compare** these to a place:
   a. further north or south.
   b. on the opposite side of Australia.

9.3 Earth's revolution around the Sun

BY THE END OF THIS MODULE, YOU WILL BE ABLE TO:
- ✓ state the time it takes Earth to orbit around the Sun
- ✓ explain why the orbital time depends on the distance of Earth from the Sun.

GET THINKING

Scan the key words in this module and make a crossword or find-a-word puzzle with these terms to share with other students in your class.

revolution
the path an object travels as it moves around another object

period
the time it takes for a satellite to complete one orbit or revolution

solar year
the time it takes for a planet to revolve once around the Sun

In Module 9.1, you learned that Earth moves about the Sun. The term **revolution** describes the orbital motion of a satellite, such as Earth, around a large body, such as the Sun. When Earth orbits the Sun, we say Earth revolves around the Sun, just as the Moon revolves around Earth. The time it takes for a satellite to complete one orbit or revolution is called the **period**. The Moon's period is 27.3 days because it takes 27.3 days to make one revolution around Earth. When the satellite is a planet orbiting the Sun, the period is called a **solar year**. Therefore, the period and solar year for Earth is 365.24 days because this is how long it takes to make one revolution around the Sun (see Figure 9.3.1).

Interactive resource
Crossword: Moving Earth

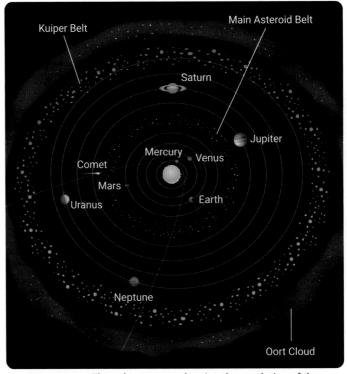

▲ **FIGURE 9.3.1** The solar system, showing the revolution of the planets around the Sun.

orbital plane
a surface that contains the orbit of a body

ecliptic
the path travelled by the Sun and planets as seen from Earth

Other planets orbit the Sun in a similar way to Earth. If you were to connect the centre of the Sun to Earth, over a year the line would sweep out a flat disc-shaped surface (Figure 9.3.2). The imaginary surface that the orbit lies in is called the **orbital plane**. All the planets of our solar system orbit with similar orbital planes to Earth. Because Earth's axis is tilted to the orbital plane, the planets and the Sun seem to travel along a path across the sky. This path is called the **ecliptic**.

▲ FIGURE 9.3.2 The planets orbit the Sun with similar orbital planes.

The time for one revolution is different for each planet. Table 9.3.1 shows how the time for a revolution (hence, a solar year) increases as the distance from the Sun to a planet increases. This distance is measured in **astronomical units** (AU), where one astronomical unit is the average distance of Earth from the Sun. This is equivalent to approximately 150 million kilometres.

The closer an object is to the Sun, the shorter its orbit and the faster it needs to travel to stay in orbit. If it were to slow down, it would spiral into the Sun. Mercury is the closest to the Sun and has the greatest speed at 47.9 km s^{-1}. Neptune is the outermost planet and travels the furthest at a leisurely 5.4 km s^{-1}. Earth, travelling closer to the Sun, moves at 29.8 km s^{-1}. The combination of greater speed and less distance results in a shorter period for planets closer to the Sun.

astronomical unit
the average distance from the centre of Earth to the centre of the Sun, equivalent to 149.6 million kilometres (AU)

▼ TABLE 9.3.1 The rotational period and distance from the Sun of the planets

Planet	Distance from the Sun (AU)	Time for one revolution (solar year)
Mercury	0.39	0.24
Venus	0.72	0.62
Earth	1	1
Mars	1.52	1.88
Jupiter	5.20	11.9
Saturn	9.58	29.4
Uranus	19.2	83.7
Neptune	30.0	163.7

ACTIVITY

Orbit calculations

1 Vesta is an asteroid that orbits the Sun in the asteroid belt between Mars and Jupiter. It has an average diameter of 525 kilometres. Can you predict the rotational period of Vesta?

▲ FIGURE 9.3.3 Vesta

2 Phobos and Deimos are the Moons of Mars. Deimos orbits Mars 2.5 times further away than Phobos. Which moon has the shortest orbit?

▲ FIGURE 9.3.4 Phobos and Deimos

9.3 LEARNING CHECK

1 **Define** 'revolution'.
2 **State** how long it takes Earth to complete one revolution of the Sun.
3 **Define** 'period'.
4 If Earth orbited closer to the Sun, **explain** how the length of a year would be different.
5 Which planet has:
   a the smallest orbital period?
   b an orbital period closest to Earth's rotational period?
6 **Explain** the relationship between a planet's orbital period and its distance from the Sun.
7 Graph the rotational period and distance from the Sun in Table 9.3.1 on a piece of graph paper with a curved line of best fit. What do you notice about the periods of the inner planets (Mercury to Mars) compared with the periods of the outer gas giants (Jupiter to Neptune)?

9.4 Seasons

BY THE END OF THIS MODULE, YOU WILL BE ABLE TO:
✓ describe seasons and how they vary in different parts of Australia.

GET THINKING

What is your favourite season? Why do you like it? In this module, we will look at the relationship between the tilt of Earth, heating and seasons. Write down what you think the source of heating is and why we have different seasons.

Video activity
Why does Earth have seasons?

The nature of seasons

The weather, length of days and the behaviour of living things change during a year. A period of time characterised by particular weather and day length is called a **season**. In southern Australia, people often recognise four seasons: spring, summer, autumn and winter (Figure 9.4.1a) Tropical northern Australia experiences a dry and wet season, which have very different characteristics from seasons in southern areas of Australia (Figure 9.4.1b).

season
a period of time characterised by weather and day length

▲ **FIGURE 9.4.1** **(a)** Autumn in southern Australia may be colourful because of the changing leaf colours of trees that have been imported from other parts of the world. **(b)** Rain and thunderstorms are a feature of the wet season in northern Australia.

Why seasons occur

Seasons result from the tilt of Earth's rotational axis. Earth's axis is tilted at 23.5° to the vertical. This means that the sunlight and heating by the Sun are different in the northern and southern hemispheres. If there was no tilt to Earth's axis of rotation, both hemispheres would be warmed equally. The Sun would always appear to be above the equator and the surface would receive the same amount of light and heat each day. There would be no seasons. However, this is not the case.

As Earth orbits the Sun, the axis continues to point in the same direction (Figure 9.4.2). During an Australian summer, the axis in the southern hemisphere is pointing towards the Sun. Six months later, in winter, the axis is pointing away from the Sun. During spring and autumn, Earth's axis lines up with Earth's direction of travel and so both hemispheres receive similar amounts of light.

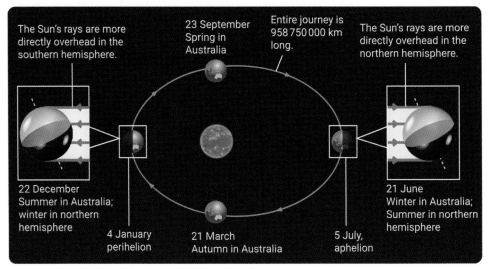

▲ FIGURE 9.4.2   Seasons are caused by Earth's tilt.

During summer, the Sun appears higher in the sky and sunlight reaches Earth nearly at a right angle to Earth's surface. This means the Sun's rays travel a shorter distance through the atmosphere and are concentrated over a smaller area (Figure 9.4.3). The more concentrated light and longer days heats the surface more than in winter.

In winter, the Sun's incoming rays are at more of an acute angle. The light travels a longer distance through the atmosphere and the sunlight is spread over a larger area. The less concentrated light and shorter days means the surface warms less than in summer.

During spring and autumn neither hemisphere is tilted towards the Sun. Therefore, the weather conditions are between the cold of winter and the warmth of summer.

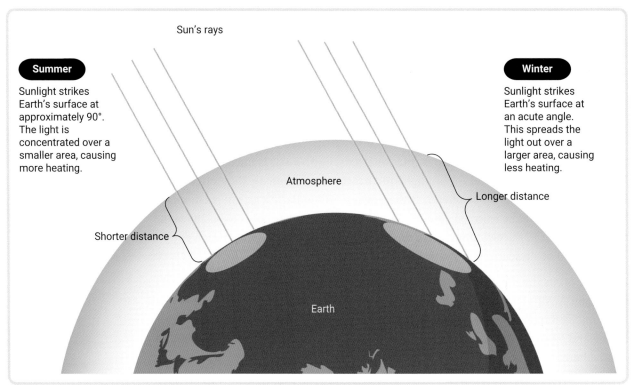

▲ FIGURE 9.4.3   The angle of sunlight affects the area lit and the amount of heating.

In northern Australia, there is less variation in the angle of the Sun's incoming rays throughout the year. This means that the seasons are not as varied as they are further south. From May to September, there is less rainfall and the land dries. This is known as the dry season. From October to April, moist air flows inland from the sea; the air rises to form clouds and rain is produced. This is known as the wet season.

ACTIVITY

The longest day of summer

The summer solstice is the day in the year when the Sun travels the longest path. It also reaches its highest point in the sky. The summer solstice occurs usually on 22 December, but it may occur a day earlier or later. The summer solstice is the longest day of the year.

Something special occurs on the summer solstice at a line of latitude called the Tropic of Capricorn. When the Sun is at its highest, it casts no shadow – it is directly overhead.

Table 9.4.1 shows information about places on Australia's east coast and in Antarctica on 22 December near midday.

Table 9.4.1 The length of shadows at different places near midday on 22 December

City or place	Latitude (degrees south of the Equator)	Angle of the Sun above Earth's surface (degrees)	Length of the day (hours)	Shadow length of a vertical 1 m stick (m)
Rockhampton	23.38 (Tropic of Capricorn)	90.0	13.6	0
Brisbane	27.47	85.4	13.9	0.08
Sydney	33.87	73.7	14.4	0.29
Melbourne	37.82	67.3	14.8	0.42
Hobart	42.88	65.4	15.4	0.46
East Antarctica	86.10	32.3	24.0	1.56

Questions
1 What is the relationship between a place's latitude and the shadow cast by the stick?
2 What happens to the length of the day as latitude increases towards the south?
3 How would the shadow lengths be different during winter?

9.4 LEARNING CHECK

1 **Define** 'season'.
2 **List** the seasons that occur where you live.
3 **List** the four seasons, in order, experienced by people in southern Australia.
4 **Describe** two ways the tilt of Earth leads to seasons.
5 Draw a labelled diagram to show Earth's orientation when it is:
   a summer in the southern hemisphere.
   b autumn in the southern hemisphere.
6 **Describe** what the world would be like if, like the planet Uranus, the axis always pointed towards the Sun.

9.5 Phases of the Moon

BY THE END OF THIS MODULE, YOU WILL BE ABLE TO:
- ✓ identify and explain the phases of the Moon.

Interactive resource
Match: Phases of the Moon

GET THINKING

Why does the Moon shine more brightly on some nights than on others? What causes the amount of moonlight to change? Over the course of a month, take photos of the Moon on different nights. Describe all the changes you see as the Moon changes its appearance.

The Moon's appearance changes over a month

We see the Moon because of reflected light. When light falls on the Moon's surface from the Sun, it is reflected towards Earth, and we see a bright surface. Earth also reflects light. Some of the reflected light is reflected again from the dark areas of the Moon not lit by the Sun, so we see a dim outline of these areas. As the Moon orbits Earth, we see different amounts of the Moon's surface that is reflecting light from the Sun (Figure 9.5.1). The amount of the Moon shining with reflected light waxes (grows) and then wanes (shrinks). This is because the angle at which we view the Moon changes.

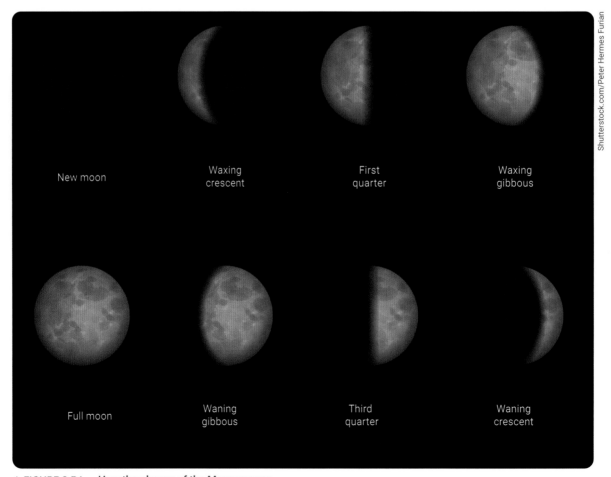

▲ FIGURE 9.5.1 How the phases of the Moon appear

Phases of the Moon

9.5

The phases of the Moon are the different appearances of Moon. It takes 29.5 days to move through the eight phases. This is slightly different from the time it takes the Moon to orbit Earth (27.3 days) because Earth has also moved in this time, and it takes 2.2 days for the Moon to reach the position for the original phase.

The phases are usually ordered beginning with the new moon (Figure 9.5.2). A **new moon** occurs when the Moon is between Earth and the Sun. When this happens, the surface of the Moon we see does not receive light from the Sun and it appears as a dark disc.

The bright area that we can see increases as the Moon is **waxing**. A **crescent** phase occurs when less than half of the side of the Moon facing Earth is lit. The bright area is thicker in the middle and tapering to points.

When we see half the Moon lit by the Sun, the Moon is in its first quarter. The quarter refers to it being a quarter way through its phases – we can actually see half the lit face.

When more than half the Moon that we see is bright, we call the phase a **gibbous** moon. The amount of the bright side we see continues to grow until the whole side of the Moon reflecting light from the Sun is visible. This is a **full moon**.

After reaching the full moon phase, the shadowed area of the Moon starts to grow, and the bright area shrinks. The bright area is said to be **waning**. Initially, this is a waning gibbous moon, with more than half of the Moon that we see lit. When the Moon has half of its lit surface visible from Earth, it has reached its third quarter. Because less than half of the side of the Moon that we see is lit, another crescent moon occurs. This is the waning crescent moon.

new moon
a phase of the Moon when the part of the Moon facing Earth is in darkness

waxing
when the bright surface area of the Moon visible from Earth is increasing

crescent
a phase of the Moon when only a small arc-shaped section of the Moon is visible from Earth

gibbous
a phase of the Moon when more than half of the illuminated face of the Moon is visible from Earth

full moon
a phase of the Moon when the whole lit face of the Moon is visible from Earth

waning
when the bright surface area of the Moon visible from Earth is decreasing

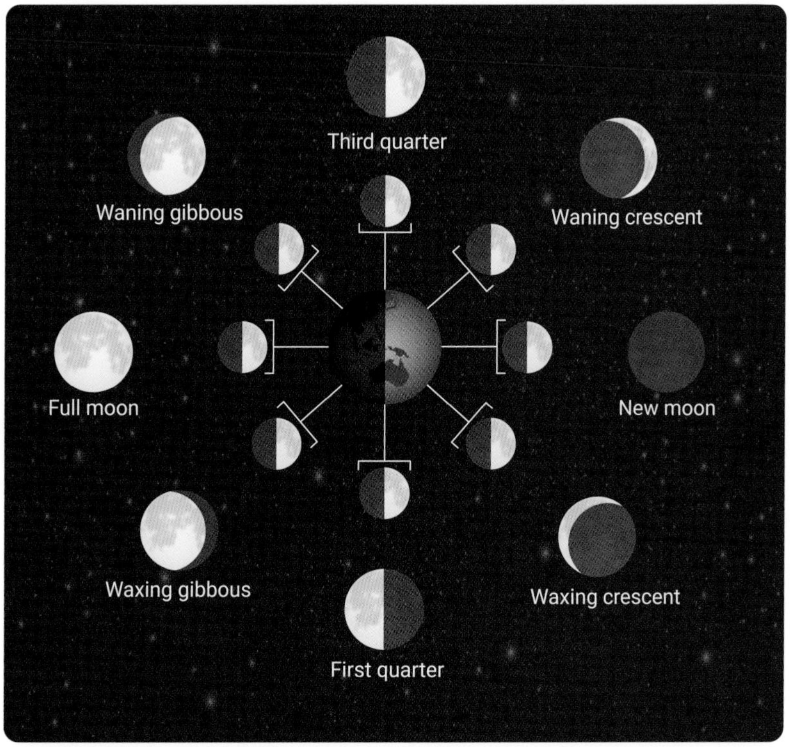

▲ **FIGURE 9.5.2** The phases of the Moon. The Moon is lit from the same direction, and we always see the same face of the Moon, but we see different amounts of its illuminated surface.

The upside-down Moon

The same phases of the Moon are seen in both the northern and the southern hemispheres, but in the north, things appear upside down and back to front compared with what we see in the south (Figure 9.5.3). In the southern hemisphere, the Moon increases in brightness from the left. In the northern hemisphere, the brightness increases from the right. This is due to the direction from which we are viewing the Moon – people in different hemispheres see the Moon from opposite sides. If you look at the same thing, such as the Moon, from opposite sides, what is on the left from one side will be on the right from the other side.

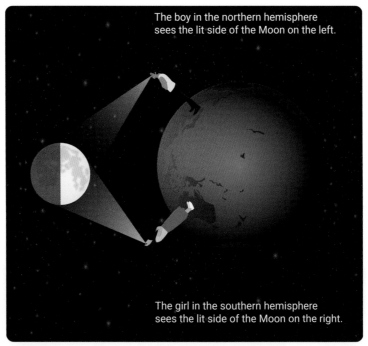

▲ FIGURE 9.5.3   What you see depends on where you are looking from.

9.5 LEARNING CHECK

1. **State** how long it takes the Moon to revolve once around Earth.
2. **Compare** the meanings of the words 'waxing' and 'waning'.
3. **List** the eight phases of the Moon, in order, starting with a new moon.
4. **Draw** a diagram to show the difference between a crescent and a gibbous phase of the Moon.
5. **Describe** the appearance of a crescent moon.
6. **Describe** how a waxing gibbous moon is different from a waning crescent moon.
7. **Create** a labelled diagram to explain the appearance of the Moon at the first quarter.

9.6 Eclipses

BY THE END OF THIS MODULE, YOU WILL BE ABLE TO:
- ✓ describe solar and lunar eclipses and explain why they occur
- ✓ explain why total and partial eclipses occur
- ✓ model how eclipses are caused.

GET THINKING

Look carefully at the images in this module. Use a torch (for the Sun), a tennis ball (for Earth) and a ping pong ball (for the Moon) to model the arrangements for each type of eclipse. Make a summary of what you think is happening.

Video activity
What are eclipses?

Extra science investigation
Modelling eclipses

Why eclipses occur

An **eclipse** occurs when light from the Sun is blocked by another object, which casts a shadow. If the Moon casts a shadow on Earth, it has blocked light from the Sun. When Earth casts a shadow on the Moon, the brightness of the Moon dims because there is less light from the Sun available to be reflected.

eclipse
when light from the Sun is blocked by another object

▲ FIGURE 9.6.1 This time-lapse photograph shows an eclipse of the Moon.

Lunar eclipses

A **lunar eclipse** occurs when the Moon enters the shadow cast by Earth (Figure 9.6.2). This happens when Earth is directly between the Sun and Moon. Therefore, a lunar eclipse occurs when the Moon is full.

As the Moon enters Earth's shadow, its surface darkens. The area where light from the Sun is completely blocked is called the **umbra**. When the Moon is in the umbra, a **total lunar eclipse** occurs. During a total lunar eclipse, the Moon may become slightly red (Figure 9.6.3). This is because the light that reaches the Moon has passed through Earth's atmosphere. In the atmosphere, blue light is scattered so mainly red light reaches, and is reflected from, the Moon.

lunar eclipse
when Earth blocks the Sun's light from reaching the Moon

umbra
the innermost, darkest part of a shadow where the light is completely blocked by an object

total lunar eclipse
when the Moon is in the umbra of Earth's shadow and none of the Moon is visible

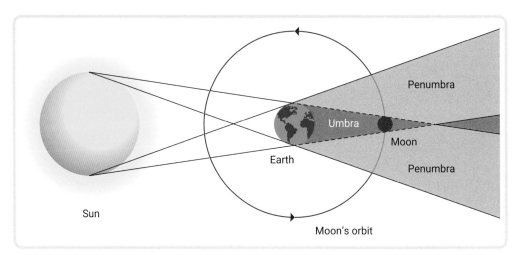

FIGURE 9.6.2 A lunar eclipse is caused by Earth moving between the Moon and the Sun.

penumbra
the outermost part of a shadow where only some of the light is blocked by an object

partial lunar eclipse
when part of the light from the Sun reaching the Moon is blocked by Earth; only a portion of the Moon is visible from Earth

Surrounding the umbra is an area where only part of the Sun's light is blocked by Earth. This is called the **penumbra**. When the Moon is partly in the umbra and partly in the penumbra, a **partial lunar eclipse** occurs. This is seen as part of the Moon being covered by a shadow.

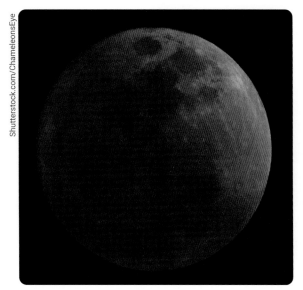

FIGURE 9.6.3 A total lunar eclipse

FIGURE 9.6.4 A partial lunar eclipse

Solar eclipses

A **solar eclipse** occurs when the Moon is between the Sun and Earth (Figure 9.6.5). Therefore, they occur during the new moon phase. As in a lunar eclipse, an umbra and penumbra occur in solar eclipses. When the Moon is between the Sun and Earth, it casts a shadow on Earth's surface, creating an umbra and penumbra, as seen in Figure 9.6.5.

solar eclipse
when the Moon passes between Earth and the Sun, blocking the view of the Sun from Earth

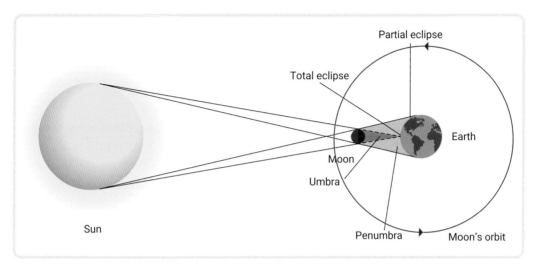

▲ FIGURE 9.6.5   A solar eclipse is caused by the Moon moving between Earth and the Sun.

Looking from Earth in the umbra, the face of the Moon completely covers the Sun, causing a **total solar eclipse**. Only the bright outer atmosphere of the Sun, the **corona**, is visible during a total solar eclipse (Figure 9.6.6). Total solar eclipses are rare though because the area of the umbra is very small on Earth's surface.

A **partial solar eclipse** is where only part of the Sun is blocked by the Moon. On Earth, a partial eclipse is seen by a person standing in the penumbra. The part of the Sun in the shadow is dark and the rest is bright.

total solar eclipse
an eclipse where Earth is in the umbra of the Moon's shadow; none of the Sun is visible from Earth

corona
the bright outer atmosphere of the Sun

partial solar eclipse
an eclipse where the Moon's shadow stops a portion of the Sun's rays reaching Earth; only a portion of the Sun is visible from Earth

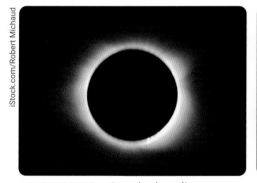

▲ FIGURE 9.6.6   A total solar eclipse

▲ FIGURE 9.6.7   A partial solar eclipse

When do eclipses occur?

The probability of an eclipse occurring depends on the:
- size of the object blocking the light
- distance between the Sun, Earth and the Moon
- inclination of orbits.

Eclipses will only occur when the Sun, Earth and the Moon are in alignment. The Moon's orbit is at a 5° angle to a line between the Sun and Earth (Figure 9.6.8). Therefore, the Moon will not be in line with the Sun and Earth at every new moon or full moon. When this happens, Earth's shadow doesn't fall on the Moon, or the Moon's shadow on Earth.

Solar eclipses occur 2–5 times a year, and there are two total solar eclipses every 3 years. Each solar eclipse only lasts a few minutes before the shadow has moved. They are only seen from a small part of Earth several hundred kilometres across because the shadow from the Moon only covers a small area. Sometimes this area is in the ocean or an uninhabited location, making solar eclipses appear rarer than they actually are.

There are between zero and three lunar eclipses every year. Each one lasts for a few hours and can be seen from anywhere on Earth.

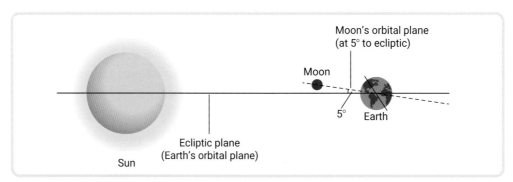

▲ FIGURE 9.6.8   The Moon orbits at an angle of 5° to the orbital plane of Earth around the Sun.

9.6 LEARNING CHECK

1. **Define** 'eclipse'.
2. **Explain** how a solar eclipse is different from a lunar eclipse.
3. Make a labelled drawing of the alignment of the Sun, Earth and the Moon needed to create a:
   a solar eclipse.
   b lunar eclipse.
4. **Explain** why both full and partial eclipses occur.
5. Use a light and two different-sized balls to model how a total and a partial eclipse are created.

9.7 Tides

BY THE END OF THIS MODULE, YOU WILL BE ABLE TO:
- ✓ describe the nature of high, low, spring and neap tides
- ✓ explain why the height of tides changes daily and over a lunar cycle.

GET THINKING

Have you ever been to the beach and noticed that the distance water moves up the beach changes during the day? The maximum distance the water travels up the beach, and when it happens, changes from day to day. Why does this happen? Write down your theory.

Video activity
The Moon and spring tides

Extra science investigation
The effect of the Moon on tides

What are tides?

Tides are the regular rise and fall of the surface of the ocean. At the beach and along the coast, the average height of the sea, and how far it reaches up the beach, changes each day. The highest level of the ocean is called **high tide** and the lowest level is called **low tide**. The difference in height of high and low tides is called the **tidal range**.

Tides do not occur at the same time every day and they vary in range. Most places have two high tides and two low tides each day. However, in some places, such as the Gulf of Carpentaria, the Gulf of Thailand, the Persian Gulf and the Gulf of Mexico, there are only one high tide and one low tide in a day.

tide
the regular rise and fall of the surface of the ocean

high tide
when the tide reaches its maximum level

low tide
when the tide reaches its lowest level

tidal range
the difference in the height of high and low tide

▲ FIGURE 9.7.1 Low tide. How do animals adapt to changing sea levels?

▲ FIGURE 9.7.2 The Bay of Fundy in Canada has the greatest tidal range in the world. How high does the water rise in the image?

The cause of tides

Tides are caused by gravity. The ocean is attracted by gravitational forces to the Moon and the Sun (Figure 9.7.3). The Moon has a greater effect on the tides because it is much closer to Earth than the Sun is. The size and direction of the forces vary around Earth. Ocean water flows towards the area of the ocean where the gravitational attraction is greatest. The gravitational pull creates two slight bulges on opposite sides of Earth – one closest to the Moon and the other on the opposite side of Earth (Figure 9.7.3). At right angles to the bulges, the water depth is lower because of the lack of gravitational force in that direction.

Tides are created as Earth rotates under the bulges. As Earth rotates, a bulge approaches a coast, and the water rises to create a high tide. As the bulge moves away, the water falls to create a low tide.

Tides do not occur at the same time every day. Earth rotates faster than the Moon orbits Earth. As Earth spins in the same direction the Moon orbits, it takes an extra 50 minutes for Earth to return to the same place relative to the Moon. This means that the gravitation pull from the Moon causes high and low tides at slightly different times every day.

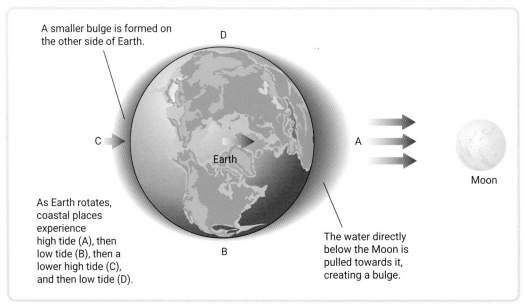

▲ FIGURE 9.7.3 The cause of tides

Spring and neap tides

When the Moon is in line with the Sun and Earth, the gravitational attraction from both the Sun and Moon is acting in the same direction. This makes the tides higher than at other times. These high tides are referred to as **spring tides** (Figure 9.7.4) and occur during the new moon and full moon phases.

spring tide
a high tide caused by the alignment of the Sun, Earth and the Moon

When the Moon is at a right angle (90°) to the line between the Sun and Earth, the Sun's gravitational attraction reduces the effect of the Moon's gravity and, therefore, the tide height. This is called a **neap tide** and occurs a week after a new or full moon. During a neap tide, there is only a small difference between high and low tides.

neap tide
a tide where the difference between high tide and low tide is small

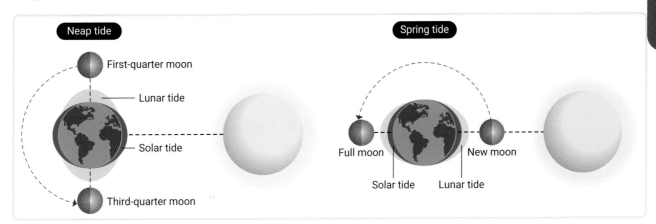

▲ **FIGURE 9.7.4** Spring and neap tides

9.7 LEARNING CHECK

1. **Define** 'tide'.
2. **Identify** how many high tides usually occur in 24 hours on the coast closest to where you live.
3. **Describe** how the Moon affects tides.
4. Which phases of the Moon occur when spring tides happen?
5. **Describe** two things that affect the size of tides.
6. **Explain** why the Sun has less effect on Earth's tides than the Moon does.
7. **Explain** why a neap tide is smaller than a spring tide.
8. **Locate** a tide chart for an area near where you live. How many minutes, on average, does the time of high tide change from day to day?

9.8 First Nations Australians' knowledge of Moon phases and tides

FIRST NATIONS SCIENCE CONTEXTS

IN THIS MODULE YOU WILL:
- explore First Nations Australians' knowledge of the phases of the Moon
- examine First Nations Australians' understanding of the relationship between the Moon and tides.

▲ FIGURE 9.8.1 In many First Nations Australians groups, the words for 'moon' and 'month' are similar; Mithaka Country, far western Queensland.

Phases of the Moon

For many thousands of years First Nations Australians have observed the night sky and built a wealth of astronomical knowledge. Their records of the repeating patterns and relationships between celestial bodies are preserved in cultural narratives (stories, songs and performances) that are used to pass on important cultural knowledge, values and beliefs), petroglyphs (images engraved into rock), stone arrangements and paintings.

Timekeeping

Prior to colonisation, timekeeping systems of many First Nations Australians groups were based on cyclical patterns of the Moon. For example, the Ngarrindjeri Peoples of the Southern Coorong district of South Australia recorded the age of children less than one year old by the number of full moons that had passed since their birth. The Takayna Peoples of north-west Tasmania applied lunar phases to the timing of gatherings; for example, the number of dark days after the Moon had disappeared.

In many First Nations Australians groups, the knowledge that the Moon takes a month to complete a cycle of the Earth is shown by the use of the same, or a closely linked, word for both 'moon' and 'month'. For example, the Meriam Mir Peoples of the eastern Torres Strait use the word *meb*, which means both moon and month. They also have names for the different lunar phases: new moon is *aketi meb*, first quarter moon is *meb degemli*, a waxing or waning moon is *eip meb* and a full moon is *giz meb*.

Evidence of First Nations Australians' understanding of the phases of the Moon is also found in petroglyphs. For example, at a site within what is now the Ku-ring-gai Chase National Park (north of Sydney) the Guringai Peoples of the Sydney region communicated their observations and understanding of the phases of the moon in a series of eight rock engravings portraying the lunar calendar.

The Moon and the Sun

First Nations Peoples have long identified relationships between the Moon and the Sun. For example, a cultural narrative of the Palawa Peoples of Tasmania tells of a Sun Man and Moon Woman who rose into the sky together on the first day. The Sun Man moved faster through the sky than the Moon Woman. To encourage her to catch up, he illuminated more of her each night until she was fully lit.

> **ACTIVITY 1**
>
> 1 **Discuss** the advantages of having shorter and longer timekeeping systems.
> 2 After reading about the cultural narrative above:
>   a **relate** the events in the narrative about the Sun Man and the Moon Woman to the phases of the Moon.
>   b **describe** what information it conveys about the relative position of the Moon and Sun at different times.
>   c **explain** the contemporary scientific understanding this demonstrates.

Relationship between the Moon and tides

Knowledge of relationships between the Moon and tides has been gained by First Nations Australians through continuous observation of the position and phases of the Moon and ocean tides. This understanding has long enabled predictions about timing and height of tides, informing many First Nations Australians' practices and technologies.

For example, First Nations Australians constructed and used fish traps. These traps were generally constructed from stone and positioned in an inter-tidal area. To be effective, their construction requires a detailed knowledge of variations in ocean currents, tide heights and times.

▲ **FIGURE 9.8.2** Stone fish traps rely on the lunar cycle: (a) fish trap on Gangalidda Country, the Gulf of Carpenteria, Queensland; (b) Toorbul Point fish trap on Ningy Ningy Country, Brisbane.

The Narungga Peoples of the Yorke Peninsula region of South Australia constructed fish traps perpendicular to the shoreline and the direction of tides and currents. The Burgiyana fish trap at Point Pearce is constructed within the minimum and maximum tidal range. At high tide the walls of the trap are submerged and water flows in. As the tide recedes, water flows out of the structure leaving fish trapped, which can then be harvested.

The Dampier Peninsula region of Western Australia experiences one of the largest tidal variations in Australia, up to 11 metres. The Bardi Peoples of this region have long used knowledge of lunar phases and the connection with tides to time safe travel between islands, during neap tides. For thousands of years the Bardi Peoples have also taken advantage of low tides to access large intertidal reefs, rock shelves and mudflats, which provide sources of food and cultural and economic resources, such as fish and pearl and trochus shells.

▲ FIGURE 9.8.3   A trochus shell

ACTIVITY 2

1. **Explain** the relationship between use of fish traps and tides.
2. **Relate** the maximum and minimum tidal range to specific phases of the Moon.
3. **Explain** the difference between a neap tide and a low tide.
4. **Suggest** why the Bardi Peoples might specifically choose to travel at neap tides rather than just low tides.

9.9 Galileo's testing of scientific ideas

SCIENCE AS A HUMAN ENDEAVOUR

BY THE END OF THIS MODULE, YOU WILL BE ABLE TO:
✓ explain how the personal beliefs of a scientist may influence the questions they choose to pursue and how they investigate those questions.

Galileo's legacy

Galileo Galilei was born in Italy in 1564. He made many scientific discoveries about how things move, gravity and the strength of materials. He investigated practical things such as the telescope, pendulum and thermometer. He was the first person to use a telescope for astronomy, finding four moons of Jupiter, Saturn's rings, and sunspots. Galileo also investigated the nature of tides. Famous scientists such as Albert Einstein and Stephen Hawking described Galileo as doing more than anyone else to pioneer modern science as we know it today.

In 1633, Galileo was sentenced to prison for arguing that Earth was not the centre of the Universe, but that it orbited the Sun. He spent the last 9 years of his life restricted to his home and was prevented from publishing any more books.

▲ FIGURE 9.9.1 Galileo Galilei made many scientific discoveries.

In Galileo's time, most educated people believed that the Sun, planets and the Moon orbited Earth (Figure 9.9.2). This was the view of the famous Greek philosopher Aristotle. Aristotle's ideas were based on some observations and logical thinking. Galileo strongly believed that ideas about how the world worked needed to be based on observation and the testing of hypotheses. He believed that, without testing ideas, people could not be confident that those ideas properly described the world. When he saw that Venus had phases and that Jupiter had moons, he understood that these observations contradicted Aristotle's view that Earth was the centre of the Universe. Instead, the observation supported the idea that Earth and the other planets orbited the Sun.

▲ FIGURE 9.9.2 Aristotle's model of the Universe. It is in Latin. Can you identify the Moon and planets?

9.9 LEARNING CHECK

1. In Galileo's time, what was society's view of the orbits of the Sun and planets?
2. How did Galileo think that ideas about how the world worked should be investigated?
3. Today, we agree with Galileo's ideas about our solar system. How do you think his findings and arguments helped to change people's ideas in his time?
4. **Research** the heliocentric model of Nicolaus Copernicus. Is it similar to today's model of the solar system? How is it different?

Quiz
Who was Galileo?

SCIENCE INVESTIGATIONS

9.10 Modelling data

SCIENCE SKILLS IN FOCUS

IN THIS MODULE, YOU WILL FOCUS ON LEARNING AND IMPROVING THESE SKILLS:

- form valid predictions based on observations and the use of scientific models
- identify patterns and relationships in scientific data.

A scientific model is a simplified representation of something complicated.

Scientific models can be:
- a physical model made with materials, such as a model of the Moon circling Earth
- mental models, such as ideas like the particle model of matter
- mathematical models, with equations and graphs to show relationships, such as a weather simulation run on a computer.

What makes a good scientific model?
- It explains how something works.
- It allows predictions to be made.
- It allows patterns and relationships to be identified and described.

Are there problems with models?
- Scientific models are only as good as the information used to build them. They often use approximations and lack some details. Models need to be tested to see how closely they represent the thing being studied, and modified to be improved as more information is acquired. For example, our weather forecasts are created using mathematical models in computers. The actual weather is used to modify the models to improve the accuracy of the forecasts.

Video
Science skills in a minute: Modelling data

Science skills resource
Science skills in practice: Modelling data

INVESTIGATION 1: MODELLING THE BEHAVIOUR OF JUPITER'S MOONS: A MATHEMATICAL MODEL

At least 79 moons orbit Jupiter. Some moons have orbital periods measured in hours while others take almost three Earth years to orbit Jupiter.

AIM

To model the relationship between a moon's distance from Jupiter and its orbital period.

YOU NEED

- ☑ access to the Internet
- ☑ a spreadsheet app, if possible
- ☑ a sheet of graph paper

WHAT TO DO

1. Make a table similar to the table on the next page.
2. Research the following moons to complete the table: Adrastea, Amalthea, Ananke, Callisto, Europa, Ganymede, Io and Lysithea.
3. Plot a graph of the orbital period (vertical axis) against each moon's distance from Jupiter (horizontal axis). Draw a curved line through the points.

RESULTS

Your table should look similar to the example table on the next page.

Remember to add enough rows for eight moons.

WHAT DO YOU THINK?

1. **Describe** the relationship between orbital period and distance from Jupiter.
2. Himalia is another moon of Jupiter, with a period of 251 Earth days.
   a. Use your line of best fit to **predict** how far from Jupiter Himalia orbits.
   b. **Research** the actual orbital distance of Himalia. How close was your prediction?
3. **Evaluate** whether the graph has the properties of a good scientific model.

Results table

Moon name	Diameter of the moon (km)	Distance from Jupiter (millions of km)	Orbital period (Earth days)	Year of discovery

CONCLUSION

Describe the relationship between a moon's distance from Jupiter and its orbital period.

INVESTIGATION 2: EARTH'S TILT AND ILLUMINATION: A PHYSICAL MODEL

AIM

To use a physical model to explore how the angle of light illumination affects surface heating

YOU NEED

- ☑ large ball
- ☑ torch
- ☑ cardboard tube about 10 cm long
- ☑ marker pens
- ☑ 30 cm ruler

WHAT TO DO

1. Mark a point on the top of the ball with a marker pen. Label this point 'equator'.
2. Mark a point 3 cm from the top of the ball. Label this point 'high latitude'.
3. Hold the tube 5 cm above the 'equator'. Shine light from the torch down the tube. Draw a line around the area lit by the torch on the ball.
4. Repeat step 3, holding the tube above the 'high latitude' point. Remember to keep the tube vertical and the same distance from the point as you did for step 3.
5. Calculate the area of the circles you have drawn and record them in your results.

RESULTS

Compare the area lit up on the ball for the 'equator' and at 'high latitude' points.

WHAT DO YOU THINK?

1. Were the areas the same size?
2. If light carries energy that turns into heat, which area will warm faster? Why do you think so?
3. If Earth's axis was vertical, would the hemispheres show different heating rates? Explain your answer.
4. Why does a tilt mean one hemisphere warms more?

CONCLUSION

How does the tilt of Earth affect heating of the hemispheres?

INVESTIGATION 3: MODELLING PHASES OF THE MOON: A PHYSICAL MODEL

AIM

To model how the phases of the Moon are created by the position of the Moon relative to Earth and the Sun

YOU NEED

- two people: an observer and a Moon carrier
- chair in the centre of the room with 2 m of clear space around it
- large ball with one half light and the other dark

WHAT TO DO

1. The observer sits in the chair.
2. The Moon carrier holds the ball in front of the observer at head height. Make sure the light half of the ball is facing the front of the room and is about 1.5 m from the observer.
3. The observer draws what they see, labelling the dark and light parts of the ball.
4. Use Figure 9.5.2 (page 267) as a guide to move the ball to the other seven moon phase positions, keeping the ball's light half always pointing towards the front of the room.
5. Have the observer turn towards the ball at each position and draw the appearance of the ball.

RESULTS

Create eight labelled drawings of the ball showing the dark and light areas that you see. Name each drawing with the phase of the moon (new moon, waxing crescent etc.) as shown in Figure 9.5.2.

WHAT DO YOU THINK?

1. Do your drawings resemble the phases of the Moon?
2. How well does the model help you to understand how the phases of the Moon are created?

CONCLUSION

Describe how a model helps our understanding of the phases of the Moon.

9 REVIEW

REMEMBERING

1 **Recall** the time it takes for:
  a Earth to orbit the Sun.
  b the Moon to orbit Earth.
  c the Moon to rotate once.
  d the rotational period of Earth.

2 **State** the angle of tilt of Earth's axis.

3 **Describe** the difference between a planet's rotation and revolution.

4 **Name** the season normally associated with the weather shown in each image. For each season, sketch the position of the Sun and Earth at this time of year.

a

b

c

d

5 **Outline** the cause of tides.

6 **List** the phases of the Moon shown below in order, starting with the new moon.

UNDERSTANDING

7 **Explain** the cause of day and night.

8 **Explain** the cause of the eclipse shown in the image.

9 **Summarise** the causes of the warm conditions that occur during summer.
10 **Distinguish** between the causes of spring and neap tides.
11 **Explain** why the Moon appears upside down in the northern hemisphere compared with what we see in Australia.
12 **Describe** examples of First Nations Australians' application of their understandings of phases of the Moon and tides.

APPLYING

13 Olivia was modelling how the angle of a beam of light on a surface warmed the surface. The results from Olivia's experiment are shown in the table.

Angle of light to the surface (degrees)	Area of light (cm²)	Change in temperature after 2 minutes (°C)
90	3.5	4.3
60	6.2	1.9
30	9.1	1.3

   a **Describe** the trends shown in the data.
   b **Explain** how the results of the modelling can be used to explain the cause of the seasons.

14 **Draw** a labelled diagram to show how increasing the distance of Earth from the Sun increases the length of a year if Earth continues in orbit at the same speed.

15 **Draw** a labelled diagram to show how a lunar eclipse occurs. On your diagram, show the positions of the Moon where total and partial lunar eclipses would be seen from Earth.

16 Finn, a student in Perth, was speaking with Annika, a student in Vancouver, Canada, about the *Apollo 11* Moon landing. Finn said that the landing site was on the eastern side of the Moon, but Annika said they had a picture showing the landing site on the western side of the Moon. **Explain** why they can both be correct.

ANALYSING

17 **Compare** the cause of a high tide with the cause of a spring tide.
18 **Compare** the seasons in southern Australia with seasons in tropical northern areas of Australia.
19 What would happen to the length of a lunar cycle if:
   a the Moon moved further away from Earth and slowed down?
   b the speed of Earth's rotation increased?
   c Earth's period of revolution was slower?

EVALUATING

20 **Assess** how the seasons would change if Earth's axial tilt increased.
21 In Module 9.10, you used models to explore processes such as day and night, phases of the Moon and the effect of axial tilt on the intensity of light on hemispheres. **Justify** the use of models in helping students understand the processes affecting Earth and the Moon.
22 **Evaluate** the effect that having two moons, rather than one, would have on Earth's tides. Give reasons for your judgement.

CREATING

23 **Construct** a concept map showing the relationship between Earth, the Sun and the Moon. Include rotations, tilt, orbital planes, revolutions, phases, tides, eclipses and seasons in your concept map.
24 A flat, two-dimensional map of the Moon circling Earth would suggest that there should be two eclipses each month. **Make** a model to show how the inclination of the Moon's orbital plane means that a shadow is not always cast on Earth or the Moon during a lunar month.

BIG SCIENCE CHALLENGE PROJECT

A new planet, nicknamed Novaterra, has been discovered orbiting a star like our Sun. You have been asked to put together a presentation on what that planet might be like.

1 Connect what you've learned

In this chapter, you have learned about the origin of seasons, phases of the Moon, tides and eclipses. Make a labelled drawing that summarises the origins and causes of these things.

2 Check your thinking

The new planet has an ocean and continents like Earth. It has two moons orbiting at different distances on opposite sides of the planet. The planet has a tilt slightly larger than Earth's and spins once every 30 hours. It is slightly further away from the star than Earth is from the Sun. Think about how these differences would make life on Novaterra different from life on Earth. Are there things you need to know to predict eclipses or Moon phases on the new planet?

3 Make an action plan

For each property of the new planet mentioned in step 2, predict the effects it would have. How would the new planet be different from Earth?

▲ Exoplanets are planets that orbit stars other than our Sun.

4 Communicate

Use your knowledge and understanding to create a presentation to explain how Novaterra compares with Earth.

Glossary

abiotic factor a non-living component of an ecosystem

acceleration due to gravity the rate at which a falling object gets faster due to the force of gravity

accuracy how close a measurement is to the correct value

aerosol a type of colloid in which a solid or a liquid is mixed in a gas

air quality index a measure of the level of pollution in the air, expressed as a number from 0 (no pollution) to 500 (maximum pollution)

air resistance friction caused by an object's motion through the air

analogy a comparison

analysis the careful study of data to look for patterns and trends

anomaly something that deviates from the standard

apex predator the organism at the top of a food chain

apparatus equipment designed for a particular use

applied force a force that is applied to an object by another object

astronomical unit the average distance from the centre of Earth to the centre of the Sun, equivalent to 149.6 million kilometres (AU)

astronomy the study of objects beyond Earth, including stars, other planets and galaxies

atom the smallest part of an element that contains the properties of that element

attractive force a non-contact force that brings two objects closer together

autotroph an organism that can make its own food

axis an imaginary line that an object spins around

balanced force a force that has an equal force acting on the same object in the opposite direction

binomial nomenclature a two-word naming system for naming living things

biological control the reduction of a pest species by using natural enemies

biology the study of living things

biomass the mass of living organisms

biomass pyramid a graphical representation of the total biomass present at each trophic level of an ecosystem

biotic factor a living component of an ecosystem

blue flame the hottest flame from a Bunsen burner; blue in colour

boiling the process of changing from a liquid to a gas at the boiling point

boiling point the temperature at which all of a substance changes from a liquid to a gas

breed a group of organisms of the same species with distinctive features

buoyancy an upwards force exerted by a fluid that opposes the downwards force exerted on an immersed object

carnivore an organism that feeds solely on animals; a meat-eater

cellular respiration the process by which cells use simple sugars (e.g. glucose) to produce energy that the organism can use

centrifuge a machine that spins very fast and separates heavier substances from lighter substances

centrifuging the process of using a centrifuge to separate a mixture

characteristic a quality or feature that makes something recognisable

chemical property a property of a substance that shows how it reacts when combined with other substances

chemistry the study of the composition and properties of matter

chlorophyll a green pigment in plants that absorbs the Sun's energy; assists photosynthesis

chromatography a process used to separate mixtures on the basis of their solubility

classification grouping things according to how similar they are

colloid a mixture of two or more insoluble substances that remains evenly mixed and does not settle over time

column a vertical division in a table

community all the organisms that live together and interact

compound a pure substance whose particles are made up of two or more different atoms chemically bonded together

compress to squash something so it takes up less space

compressibility the ability to be compressed (or squashed)

concentrated has a large amount of solute in a certain volume of solution

concentration the amount of solute dissolved in a certain volume of solution, often measured in grams per litre

conclusion a judgement reached by reasoning

condensation the process of changing from a gas to a liquid

condenser the piece of equipment in a distillation apparatus that cools the gas so that it changes to liquid

consumer an organism that must consume its food; an animal; a heterotroph

contact force a force applied by one object to another object when they are touching each other

continuous data measurements on an infinite scale so any value between two numbers is possible

control test a test in an investigation in which nothing is changed

controlled variable a factor that needs to be kept the same throughout a scientific investigation so that it does not influence the result

corona the bright outer atmosphere of the Sun

crescent a phase of the Moon when only a small arc-shaped section of the Moon is visible from Earth

crystallisation the process in which excess solute in a solution forms crystals

cycle a process that recycles resources

data the numbers or observations collected during an experiment

day the time it takes for a planet to make one rotation around its axis

daytime the time of day between sunrise and sunset

decantation the process of decanting

decanting pouring off the top, less dense liquid

decomposer an organism, such as a fungus or bacteria, that breaks down dead matter

deforestation the removal of naturally occurring forest by logging or burning

density the mass of a substance in a certain volume

dependent variable the factor that may be affected by the independent variable; the factor that can be measured or counted

deposition the process of changing from a gas to a solid

detritivore an organism that feeds on dead or decaying matter

dichotomous key a tool used by scientists to classify living things; two choices are given at each level, which helps to narrow down what species something is

diffusion the movement of particles from an area of high concentration to an area of low concentration

dilute has a small amount of solute in a certain volume of solution

discrete data data where there is only a limited number of possibilities

dissolve when a substance is mixed with another and the particles from both substances spread out evenly until they are too small to see

distillate the liquid collected during the distillation process

distillation the process used to separate solutions that collects both the solute and the solvent

domesticated adapted over time to live with humans

eclipse when light from the Sun is blocked by another object

ecliptic the path travelled by the Sun and planets as seen from Earth

ecosystem the living and non-living factors of an environment and all their interactions

electrostatic force a force resulting from the electrical charge of two objects

element a pure substance made up of only one type of atom; it cannot be broken down into a simpler substance

emulsion a type of colloid in which a liquid is mixed in another liquid

endangered in danger of becoming extinct

energy pyramid a graphical representation of the total energy present at each trophic level of an ecosystem

environment a unique set of non-living and living factors for a particular area and time

environmental science the study of the conditions of the environment and their effects on all organisms

equipment tools used to perform a task

evaporating basin a small porcelain dish used to evaporate solvent from a solution

evaporation the process of changing from a liquid to a gas at a temperature lower than the boiling point

extinct no longer in existence

fair test a way of finding the answer to a question that ensures the answer is valid

fertile able to produce offspring

field a region of space in which a non-contact force exists

filter funnel a funnel used to hold filter paper during filtration

filter paper paper with very fine holes (pores) that allow only very small particles to pass through

filtering performing the process of filtration

filtrate the substance that passes through the filter paper, usually a liquid

filtration a process used to remove solid substances from a liquid or gaseous mixture based on differences in the size of particles

flocculant a chemical added to a colloid to make the particles clump together

flocculation the process in which particles in a colloid join to form larger clumps

flow move from one place to another in a steady stream

foam a type of colloid in which a gas is mixed in a liquid

food chain a single linear diagram that shows the way energy and matter are transferred from producer to consumers

food web a group of interlinked food chains that gives an overall picture of how energy and matter are transferred through an ecosystem

force a push, pull, twist or squeeze experienced by an object when it interacts with another object

force arrow an arrow drawn on a diagram to illustrate the direction and relative strength of a force

force of attraction a force that pulls objects together

freezing the process of changing from a liquid to a solid

full moon a phase of the Moon when the whole lit face of the Moon is visible from Earth

friction a force that acts against the direction of motion, or intended motion, of an object because of an interaction between their surfaces

fulcrum the point at which a lever is supported and rotates

gas a state of matter in which the particles are very far apart and move with a lot of energy

gel a type of colloid in which a solid is mixed in a liquid

geology the study of the liquid and solid parts of Earth

gibbous a phase of the Moon when more than half of the illuminated face of the Moon is visible from Earth

gravitational force a force that results from the mass of an object and acts on other objects

gravity a force applied by one mass on another mass

hazard something that has the potential to harm

herbivore an organism that feeds solely on plants; a primary consumer

heterotroph an organism that cannot make its own food and so gains nutrition by ingesting other sources

hierarchical an order of importance

high tide when the tide reaches its maximum level

humidity the amount of water vapour in the air

hybrid the offspring of a mating between two different species

hypothesis a testable explanation for something based on existing knowledge; a testable statement of the predicted relationship between the independent and dependent variables

immiscible unable to mix; separates into layers if combined

inclined plane a sloping ramp

independent variable the factor that you choose to vary in your investigation

inference a reasonable conclusion based on observations

insoluble cannot dissolve in another substance

interaction an action that occurs as two objects have an effect on each other

introduced species a species that was not part of the original ecosystem; for example, plants, animals and micro-organisms brought into Australia from other countries

invasive species an introduced species that disrupts the ecosystem

kinetic energy the energy of an object due to its motion

length the distance between two points, measured in metres (m), centimetres (cm) or millimetres (mm)

lever a solid plank or bar that rotates about a point

linked key (tabular key) a descriptive dichotomous key made of numbered questions or statements

Linnaean classification system a classification system consisting of a hierarchy of groups, with each group being further divided into smaller groups based on similar characteristics

liquefication the process of changing into a liquid

liquid a state of matter in which the particles are close together but unable to break free of each other

low tide when the tide reaches its lowest level

lunar eclipse when Earth blocks the Sun's light from reaching the Moon

lustre how shiny a metal is

magnet a material that produces a magnetic field

magnetic force a force acting between magnetic poles

magnetism a force that is experienced by metals such as iron

magnify to make something appear larger

mass the amount of matter in an object, measured in kilograms (kg), grams (g) or milligrams (mg)

matter anything that takes up space and has mass

mean the calculated 'central' value of a set of number; an average

mechanical advantage a measure of the force multiplication provided by a machine

melting the process of changing from a solid to a liquid

melting point the temperature at which a substance changes from a solid to a liquid

meniscus the curved surface of a liquid when it is in a thin tube

method the steps that were taken during a scientific investigation; written in past tense

mixture a substance made up of different types of particles that are physically combined

mnemonic a memory aid that uses the pattern of letters in words

model a simplified explanation that makes something easier to explain or understand

moon a natural satellite of a planet

motion the change in position of an object over time

multicellular composed of many cells

native species an organism that originated and developed in the environment

neap tide a tide where the difference between high tide and low tide is small

net force the sum of all forces acting on an object; also known as total force

new moon a phase of the Moon when the part of the Moon facing Earth is in darkness

newton the unit of force (N)

night-time the time of day between sunset and sunrise

non-contact force a force that an object exerts on another object without the objects touching each other

numbers pyramid a graphical representation of the total number of organisms at each trophic level of an ecosystem

objective not influenced by personal feelings or opinions

observation data collected through the senses (sight, smell, taste, touch or hearing) or with measuring tools

omnivore an organism that eats both plants and animals

opaque cannot be seen through

orbit the regular, repeating path an object takes in space around another object

orbital plane a surface that contains the orbit of a body

organism a living thing

parallax error an error in the reading of an instrument due to the viewing angle

partial lunar eclipse when part of the light from the Sun reaching the Moon is blocked by Earth; only a portion of the Moon is visible from Earth

partial solar eclipse an eclipse where the Moon's shadow stops a portion of the Sun's rays reaching Earth; only a portion of the Sun is visible from Earth

particle a tiny unit of matter

particle theory of matter a theory that states that all matter is made up of particles that are in constant motion

pattern when data repeats in a predictable manner

penumbra the outermost part of a shadow where only some of the light is blocked by an object

period the time it takes for a satellite to complete one orbit or revolution

photosynthesis the process by which plants use light energy from the Sun to produce simple sugars (e.g. glucose) in a series of chemical reactions

physical property a property of a substance that can be observed or examined without changing the composition of the substance

physics the study of matter, energy and the interaction between them

planet a natural body that orbits a star

pollutant a substance introduced into an environment that can be harmful

population organisms of one species living together

prediction the expected results

primary consumer (first-order consumer) an organism that eats a producer; herbivore or omnivore

procedure a set of instructions to follow; written in the present tense

producer an organism that produces its own food; usually a plant; an autotroph

property a characteristic or feature of a substance

pulley a wheel on an axle that enables a change in direction of a rope or cable

pulling force a force applied by an object away from another object

pure substance a substance made up of the same type of particle

pushing force a force applied by an object towards another object

qualitative data non-numerical information that relates to a quality, type, choice or opinion

quantitative data numerical information that is counted or measured and expressed as numbers

reaction force a force acting in the opposite direction to the applied force and on the object that exerted the applied force

reliability how similar the results of the same experiment are

repeatable the same results are obtained when the same person conducts the same experiment

reproducible the same results are obtained when a different person conducts the same experiment

repulsive force a non-contact force that pushes two objects away from each other

residue what is left in the filter paper after filtration

result the information gained from an experiment

revolution the path an object travels as it moves around another object

rotation the motion of an object around an internal axis

row a horizontal division in a table

safety flame the cooler flame from a Bunsen burner that is easily visible because it is orange; also known as the orange flame

saprophyte an organism that digests dead matter before ingesting it; also known as a saprotroph

satellite an object in orbit around a larger object

saturated solution a solution that has the maximum amount of solute dissolved in the solvent

science the study of the natural and physical world by asking questions, making predictions, gathering evidence, solving problems and revising knowledge

scientific method a systematic way of gaining knowledge

scientist a person who uses research to gain knowledge and understanding of any area of science

screw a long, inclined plane wrapped around a solid cylinder

season a period of time characterised by weather and day length

secondary consumer (second-order consumer) an organism that eats a primary consumer; carnivore or omnivore

sediment the insoluble solid that settles on the bottom of a suspension

sedimentation the process of particles settling on the bottom of the liquid part of a suspension

simple machine a device to increase the size of an applied force

solar eclipse when the Moon passes between Earth and the Sun, blocking the view of the Sun from Earth

solar year the time it takes for a planet to revolve once around the Sun

solid a state of matter in which the particles vibrate in fixed positions close to each other

solidification the process of changing from a liquid to a solid

solubility how much of a substance can dissolve in a certain volume of another substance

soluble can dissolve in another substance

solute a substance that dissolves in another substance to form a solution

solution a mixture formed when a solute dissolves in a solvent

solvent a substance that dissolves another substance to form a solution

space the three-dimensional region (length, width and height) where an object exists

species a group of similar organisms that can breed to produce fertile offspring

specific name the second part of the scientific name that identifies the species within a genus

spring balance a device for measuring force; also called a force meter or newton meter

spring tide a high tide caused by the alignment of the Sun, Earth and the Moon

state of matter one of the forms in which matter can exist – solid, liquid or gas

stereotype a set idea about something or someone

sterile cannot produce offspring

structure how something is built or organised

subjective based on personal feelings or opinions

sublimation the process of changing from a solid directly to a gas

supersaturated solution a solution that contains more solute than is normally able to dissolve in it at a certain temperature

suspend hang or keep from falling

suspension a mixture of at least one insoluble solid and a liquid or solution, where the insoluble substance settles to the bottom of the container over time

taxa groups in the classification system of organisms (singular: taxon)

taxonomy the study of naming, defining and classifying organisms

temperature how hot or cold something is, measured in degrees Celsius (°C)

tension force a force that acts to pull along a rope, cable, string, wire or chain

tertiary consumer (third-order consumer) an organism that eats a secondary consumer; carnivore or omnivore

tidal range the difference in the height of high and low tide

tide the regular rise and fall of the surface of the ocean

tilt sloping away from the vertical

time how long something takes, measured in hours (h), minutes (min) and seconds (s)

total force the sum of all forces acting on an object; also known as net force

total lunar eclipse when the Moon is in the umbra of Earth's shadow and none of the Moon is visible

total solar eclipse an eclipse where Earth is in the umbra of the Moon's shadow; none of the Sun is visible from Earth

transparent see-through

tree diagram a diagrammatical dichotomous key made by branching, which represents the splitting of each group

trend the general direction of data; how one variable affects another

trophic level the level or position in a food chain

Tyndall effect when the insoluble particles in a colloid interact with a beam of light to enable you to see the beam of light

umbra the innermost, darkest part of a shadow where the light is completely blocked by an object

unbalanced force a force that does not have an equal force acting on the same object in the opposite direction

unicellular composed of only one cell

unit a fixed quantity used as a standard of measurement

unsaturated solution a solution that can dissolve more solute

valid investigation an experiment that tests the hypothesis

validity the extent to which an investigation tests the hypothesis

vaporisation the process of forming a gas; evaporation, boiling or sublimation

variable a factor that could influence the result of an investigation

viscosity a liquid's resistance to flowing

volume the amount of space occupied, measured in litres (L) or millilitres (mL)

waning when the bright surface area of the Moon visible from Earth is decreasing

water stress when the amount of water needed is greater than the amount of water available

water vapour the gaseous form of water

waxing when the bright surface area of the Moon visible from Earth is increasing

wedge a triangular-shaped tool tapering to a thin edge that acts as a portable inclined plane

weight the downwards force on an object due to gravity

wheel and axle a solid rod connected to a wheel

work the energy transferred to or from an object when a force is continuously applied to the object over a distance as the object moves

Index

A

abiotic factor 188
acceleration due to gravity 238–40
accuracy 55
aerosols 113
air 114
air hole, Bunsen burner 20
air quality 114–15
air quality index (AQI) 114–15
air resistance 227
alloys 100
analogy 163
Animalia 167, 170
animals, classifying 175–6
anomaly 48
ape classification 177–8
apex predator 192
apparatus 16
applied force 222
Archaebacteria 169, 170
Aristotle 279
arlengarr 142
astronomical units (AU) 261
astronomy 6
astrophysics 6
atoms 102
attractive force 230
autotroph 167, 190
average, finding 49
axis 258

B

bacteria 169, 170
balance scale 225
balanced forces 232–3
banksia cone 143
barrel, Bunsen burner 20
beach water 111
beakers 16
bicornate basket 143
binomial nomenclature 171
biochemistry 6
biological control 210
biology 6
biomass 203
biomass pyramids 203–4
biotic factors 188
blue flame, Bunsen burner 20, 21
boat, producing drinking water on 88–9
boiling 85
boiling points 85, 126

branching tree 157
breeds 161
Bunsen burner 17
    flames, types of 20
    lighting 21
    parts of 20
    turning off 21
buoyancy 233

C

carnivore 192
cellular respiration 191
Celsius (°C) 25
centrifuge 127
centrifuging 127–8
characteristics 154, 155
chemical property 124
chemical substance, physical property of 124
chemistry 6
chihuahuas 161
chitin 168
chlorophyll 190
chromatography 140
classification
    changing 177–8
    defined 154
    dichotomous keys 156–8
    different species 159–61
    effectiveness of 155
    First Nations Australians' 175–6
    food 155
    Linnaean 162–6
clean water, separating techniques to obtain 142–3
cold-pressing 144
collar, Bunsen burner 20
colloids 112–13
    defined 112
    structure and properties of 112
    types of 113
colour, as physical property 124–5
column graph 50
column, table 48
communicators, scientists as 62
community 188
compounds 102
compressed 79, 81, 83
compressibility 79
concentrated 107, 108
concentration, solutions 107–8
conclusion 56
condensation 86

condenser 136
conical flasks 16
consumer 192
contact forces 226–7
continuous data 50
control test 44
controlled variables 39, 40
coolamon 142
corona 271
COVID-19 pandemic 5
crescent moon 267
crystallisation 135
crystals, making 137–8
cycle 200
cylinder, graduated/measuring 24

D

Dalton, John 72
Damara sheep 160
data
    analysis 53
    continuous 50
    defined 48
    discrete 50
    organising 48–9
    patterns/trends 50
    qualitative 8, 10, 50
    quantitative 8, 10, 49
data collection techniques 46
day 259
daytime 258
decantation 128–9
decomposers 200–201
deforestation 209
De Mestral, George 27
Democritus 72
density 125
dependent variables 39, 40
deposition 86
deserts, collecting drinking water in 88
detritivores 201
dichotomous keys 156–8
diffusion 83
dilute 107, 108
discrete data 50
dissolve 104
distances in space 257
distillate 136
distillation 136
domesticated 161
drinking water, collection of 88–9
dropper 18

E

Earth 256, 258
    axis 258
    revolution around Sun 260–1
    rotation of 259–60
    tilt 258, 259
eclipses 269–72
    defined 269
    lunar 269–70
    occurrence 272
    solar 271
ecliptic 260
ecosystems 188–9
    defined 188
    energy in 190–1
    food chains 192–4
    food webs 195–9
    human impact 209–11
    modelling with pyramids 202–5
    movement of energy and matter in 200–201
    paddock 189
electric hot plate 17
electrostatic force 228, 229
elements 102
emulsions 113
endangered species 209
energy
    in ecosystems 190–1
    flow 192, 200
    kinetic 73
    movement in ecosystem 200–201
    Sun 86, 87
energy pyramids 202
environmental science 6
environments 188
equipment 16–19
    defined 16
    heating 17
    for holding 18
    to hold liquids 16
    magnifying 17
Erlenmeyer flasks 16
evaluation 55
evaporating basin 135
evaporation 85, 134–5
extinct species 209

F

fair test 34
fertile 160
fields 228
filter funnel 132
filter paper 131
filtering 131
filtrate 131
filtration 131–3
First Nations Australians
    classification systems 175–6
    knowledge of Moon phases and tides 276–8
    separation techniques traditionally used by 142–4
    spear-throwing technology 247
    traditional ecological knowledge 206–8
first-order consumer 192
flame, Bunsen burner 20
flocculant 140
flocculation 140
flour 124
flow 79, 81, 82–3, 192, 200
foams 113
food, separating techniques to obtain 142–3
food chains 192–4
    constructing 193–4
    defined 192
    interconnecting 195
    roles in 192–3
    trophic levels 193
food webs 195–199
    creating 198
    defined 196
    drawing 197
    interpreting 197
force arrow 234
force of attraction 75
forceps 18
forces 220–49
    applied 222
    balanced 232–3
    come in pairs 222
    contact 226–7
    defined 220
    describing 224
    effect of mass 236–7
    function of 220
    measuring 224–5
    net 234–5
    non-contact 228–31
    reaction 222
    simple machines and 241–6
    unbalanced 232–3
freezing 86
friction 226–7
fulcrum 245
full moon 267
fungi 168, 170

G

Galilei, Galileo 279
gases 75–6
    defined 75
    properties of 82–3
gas hose, Bunsen burner 20
gas tap, Bunsen burner 20, 21
gels 113
geochemistry 6
geology 6
gibbous moon 267
Goodall, Dame Jane 62
Gouldian (rainbow) finch 175
graduated cylinder 24
gram 23
graphs
    drawing 52
    key features of 52
    types of 50–1
gravitational force 228
gravity 221
Great Danes 161

H

hand-picking 142
hardness 126
hazards 12
heating
    equipment 17
    substance 84–5
herbivore 192
Hero of Alexandria 248
heterotroph 167, 192
hierarchical levels 163
high tide 273
holding equipment 18
hook 27
humans 159
    classification of 164
humidity 88
hybrid species 160
hypothesis 35, 42
    testing 44–7

I

ice-cream 86–7
immiscible substances 125
inclined plane 241, 242
independent variables 39, 40
inferences 11
information, representing in diagram 202

insoluble 104, 125
interaction 221
introduced species 209–11
invasive species 209

J

jawun 143
Jupiter 256

K

kinetic energy 73
kingdoms 163, 164, 167–70
    classifying organisms into 167

L

length 25
lever 241, 245
lighting, Bunsen burner 21
line graph 51
linked key 157–58
Linnaean classification system 162–6
    defined 162
    divisions in 163–5
    modelling 166
Linnaeus, Carl 162
liquefication 84
liquids 126
    defined 75
    properties of 80–1
living things
    Linnaean classification of 162–6
    naming 170–4
loop fasteners 27
low tide 275
lunar eclipse 271–2
lustre 126

M

magnet 139, 229
magnetic force 228, 229
magnetic separation 139
magnetism 126, 139
magnify objects 17
marine biology 6
mass 23, 70
    defined 236
    effect, on force 236–7
    vs weight 238
matter
    classifying 98–101
    cycling of 200–201
    defined 70, 98
    describing 71

    movement in ecosystem 200–201
    particle theory of 72–3, 75
    states of *see* states of matter
mean 49
measurement 23–6
    force 224–5
    length 25
    mass 23
    temperature 24–5
    time 25
    volume of liquids 24
mechanical advantage 241–2
medicines, separating techniques to obtain 144
melting 84
melting points 84, 126
meniscus 24
Mercury 256
method 45
microscopes 17
mixtures
    classification of 106
    classifying matter as 99–101
    defined 98
    examples of 100
    teaching about 101
mnemonics 34
models 72, 73, 98
Monera 169
Moon 256, 257, 277
    appearance 266
    eclipse 269–70
    phases of 267–8, 276
    and tides 277–8
    upside down 268
motion 220
mule 160
mulga seeds 143
multicellular 167

N

naming living things 171–4
neap tides 275
net force 234–5
new moon 267
new species, naming 173
newtons 224
night-time 258
Nobel, Alfred 3
Nobel Prizes 3
non-contact forces 228–31
    defined 228
    effect of 230
    features of 230
    and force fields 228
    types of 228–9
numbers pyramid 204–5

O

objective 157
observations 8–10, 41
omnivores 193
opaque 112
orange flame, Bunsen burner 20, 21
orbit 256
orbital plane 260
organisms 159
    classifying into kingdoms 17
    naming 171–4

P

paddock ecosystem 189
parallax error 24
partial lunar eclipse 270
partial solar eclipse 271
particle theory of matter 72–3, 75
particles
    defined 72
    in pure substances 102
patterns 50
penumbra 270
period 260
personal protective equipment (PPE) 14–15
photosynthesis 167, 190
physical property
    boiling points 126
    of chemical substances 124
    colour 124–5
    defined 124
    density 125
    melting points 126
    soluble/insoluble 125
    transparency 125
physics 6
phytoplankton 203
planets 256
Plantae 167–8, 170
plants, classifying 176
pollutants 114
population 188
predictions 43
primary consumer 192
principles of rocket propulsion 248
procedure 44, 45
producer 192
property 71
Protista 168, 170
pulley 241, 244
pulling force 226
pure substances
    classifying matter as 99–101
    defined 98

examples of 100
particles in 102
teaching about 101
pushing force 226

Q

qualitative data 108
   defined 8, 50
   vs quantitative data 10
quantitative data
   defined 8, 49
   vs qualitative data 10
questions 41

R

reaction forces 222
recording results 48–9
recycling, separation techniques used in 145
reliability 55
repeatable 55
reproducible 55
repulsive force 230
residue 131
results 59–61
   analysing 50–4
   defined 48
   recording 48–9
retort stand 18
revolution
   defined 260
   Earth's, around Sun 260–1
river red gums, classification of 165
rockets 248
rotation of Earth 258–9
row, table 48
ruler 25

S

safety
   equipment 14–15
   hazards in laboratory 12
   rules 12–13
safety flame 20
saprophytes 201
satellite 256
saturated solution 108
scales 225
science
   branches of 6–7
   defined 4
science reports 57–61
scientific method

analysing results 50–4
asking questions 41
conclusion 56
defined 34
evaluation 55
hypothesis 42, 44–7
predictions 43
recording results 48–9
steps of 35
usage of 36
variables 38–40
scientific names 171
   comparing 172
   conventions 172
   of Australian organisms 174
scientists
   as communicators 62
   defined 4
   naming 7
   work 6–7
screw 241, 243
seasonal calendars 206–7, 279
seasons 263–5
   defined 263
   nature of 263
   occurrence 263–5
seawater 135
secondary consumer 192
second-order consumer 192
sediment 127
sedimentation 127
separation, solutions 134
   crystallisation 135
   distillation 136
   evaporation 134–5
separation, suspensions
   centrifuging 127–8
   decantation 128–9
   filtration 131–3
   sedimentation 127
separation techniques
   chromatography 140
   flocculation 140
   magnetic separation 139
   used by First Nations Australians 142–4
   used in recycling 145
simple machines 242–7
   defined 242
smell 126
Socrates 5
solar eclipses 273
solar year 262
solidification 86
solids
   defined 75
   properties of 78–9

solubility, of substance 104
soluble 125
solute 106
solutions 106–10
   appearance of 107
   concentration 107–8
   defined 106
   saturation 108
   separating *see* separation, solutions
   structure of 106–7
   transparent 107
solvent 106
space 70
spatula 18
spear-throwing technology 247
species 160
   endangered 209
   extinct 209
   introduced 209–11
   invasive 209
specific name 171
spring balance 224
spring tides 275
states of matter 74–7
   changing 84–7
   concept map for 74
   defined 74
   modelling 76–7
   summarising 77
   see also gases; liquids; solids
stereotypes 4
sterile 160
stirring rod 18
structure 163
subjective 157
sublimation 85
substances
   chemical, physical property of 124
   cooling 86
   heating 84–5
   immiscible 125
   insoluble 104
   solubility of 104
sugar 124
summer solstice 265
Sun 256, 277
   Earth's revolution around 260–1
   eclipse 271
   energy of 86, 87
supersaturated solution 108
suspend 110
suspensions 110–11
   defined 110
   separating *see* separation, suspensions
   structure and properties of 110
Systema Naturae (Linnaeus) 162

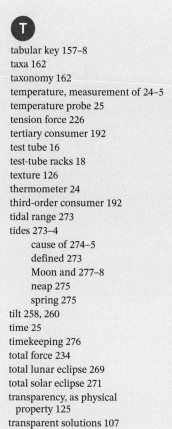

tabular key 157–8
taxa 162
taxonomy 162
temperature, measurement of 24–5
temperature probe 25
tension force 226
tertiary consumer 192
test tube 16
test-tube racks 18
texture 126
thermometer 24
third-order consumer 192
tidal range 273
tides 273–4
    cause of 274–5
    defined 273
    Moon and 277–8
    neap 275
    spring 275
tilt 258, 260
time 25
timekeeping 276
total force 234
total lunar eclipse 269
total solar eclipse 271
transparency, as physical property 125
transparent solutions 107
tree diagram 157

trends 42, 50
trophic levels 193
tug-of-war 223
tweezers 18
Tyndall effect 112

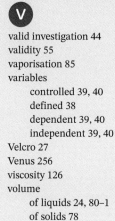

umbra 271
unbalanced forces 232–3
unicellular 167
units 224
universal solvent 107
unsaturated solution 108

V

valid investigation 44
validity 55
vaporisation 85
variables
    controlled 39, 40
    defined 38
    dependent 39, 40
    independent 39, 40
Velcro 27
Venus 256
viscosity 126
volume
    of liquids 24, 80–1
    of solids 78

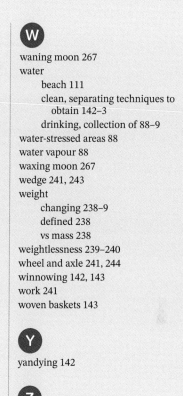

waning moon 267
water
    beach 111
    clean, separating techniques to obtain 142–3
    drinking, collection of 88–9
water-stressed areas 88
water vapour 88
waxing moon 267
wedge 241, 243
weight
    changing 238–9
    defined 238
    vs mass 238
weightlessness 239–240
wheel and axle 241, 244
winnowing 142, 143
work 241
woven baskets 143

Y

yandying 142

Z

zero gravity 239–240

Additional credits

Chapter 1
- **Page 14: Figure 1.4.5:** top row left to right: iStock.com/alacatr, Shutterstock.com/KhamkhlaiThanet, iStock.com/t_kimura; second row left to right: Shutterstock.com/Sergey Ryzhov, Shutterstock.com/Gerisima, iStock.com/shank_ali; third row left to right: Alamy Stock Photo/fStop Images GmbH, Shutterstock.com/photographyfirm, Shutterstock.com/Red Herring; bottom row left to right: Shutterstock.com/AnaLysiSStudiO, iStock.com/zilli, Shutterstock.com/Jimmy S. Aditya.
- **Page 31:** Newspix/Kym Smith, Science Photo Library.

Chapter 6
- **Page 153: Figure 6.0.1:** top row left to right: Shutterstock.com/FloridaStock, Shutterstock.com/Ramon Carretero, Shutterstock.com/Michal Pesata, iStock.com/bazilfoto; middle row left to right: Shutterstock.com/4dings, Shutterstock.com/Shane Gross, iStock.com/Bulgac, Shutterstock.com/irin-k; bottom row left to right: Shutterstock.com/Ken Griffiths, Shutterstock.com/Milan Zygmunt, Shutterstock.com/Wichai Prasomsri1, Shutterstock.com/Andrey Pavlov.
- **Page 155: Figure 6.1.2:** top row left to right: Shutterstock.com/Joe Gough, Shutterstock.com/Janet Faye Hastings, Shutterstock.com/Igor Dutina, Shutterstock.com/New Africa; bottom row left to right: Shutterstock.com/Africa Studio, iStock.com/dlerick, Shutterstock.com/oksana2010, Shutterstock.com/Popartic.
- **Page 157: Figure 6.2.2:** left to right: iStock.com/davidf, Shutterstock.com/Inked Pixels, Shutterstock.com/Cherdchai charasri, Shutterstock.com/azure1.
- **Page 174: Figure 6.6.5:** top row left to right: Alamy Stock Photo/imageBROKER, Shutterstock.com/Anne Powell, Alamy Stock Photo/Roberto Nistri; bottom row left to right: Shutterstock.com/Natalia Kuzmina, Alamy Stock Photo/Martin Fowler, iStock.com/prill.
- **Page 180: Figure 6.9.2:** top row left to right: iStock.com/kyoshino, iStock.com/volschenkh, Shutterstock.com/gloverk, Shutterstock.com/Purple Clouds; second row left to right: Shutterstock.com/Craig Walton, iShutterstock.com/kwanchai.c, Shutterstock.com/creativesunday, iStock.com/studiocasper; third row left to right: Shutterstock.com/A Bennion, Shutterstock.com/Focus and Blur, Shutterstock.com/Rabbitmindphoto, Shutterstock.com/Rabbitmindphoto; bottom row left to right: Shutterstock.com/gloverk, Shutterstock.com/PixelSquid3d, Shutterstock.com/PhotobyTawat.
- **Page 185:** clockwise from top right: Shutterstock.com/Ramon Carretero, iStock.com/bazilfoto, Shutterstock.com/Shane Gross, Shutterstock.com/irin-k, Shutterstock.com/Milan Zygmunt, iShutterstock.com/Andrey Pavlov, Shutterstock.com/Wichai Prasomsri1, Shutterstock.com/Ken Griffiths, iStock.com/Bulgac, iStock.com/4dings, Shutterstock.com/Michal Pesata, Shutterstock.com/FloridaStock.

Chapter 7
- **Page 193: Figure 7.3.2:** left to right: iStock.com/jorgeantonio, iStock.com/Benambot, Shutterstock.com/WildlifeWorld, Shutterstock.com/SJ Duran, iStock.com/DuncanSharrocks.
- **Page 195: Activity:** left to right: iStock.com/drferry, iStock.com/tracielouise, Shutterstock.com/homebredcorgi, Shutterstock.com/KarenHBlack, Shutterstock.com/imagevixen.
- **Page 195: Figure 7.4.1:** clockwise from left: Shutterstock.com/Susan Flashman, Shuttesrtock.com/Adrian Eugen Ciobaniuc, iStock.com/LisaStrachan, Shutterstock.com/Oligo22, Shutterstock.com/William Edge.
- **Page 196: Figure 7.4.2:** top: Shutterstock.com/Josh Prostejovsky; second row left to right: Shutterstock.com/Ken Griffiths, Shutterstock.com/Ger Bosma Photos; third row left to right: Shutterstock.com/7th Son Studio, Shutterstock.com/Marco Tomasini, Shutterstock.com/Yatra; fourth row left to right: Shutterstock.com/Ikhwan Ameer, Shutterstock.com/Hanahstocks, Shutterstock.com/Ramadhan Wahyu Pradana. Shutterstock.com/fritz1.
- **Page 198: Activity:** top row left to right: Shutterstock.com/OllyPlu. Shutterstock.com/nwdph. Shutterstock.com/andrepra. Forest & Kim Starr, CC BY 3.0 US, via Wikimedia Commons; second row left to right: Shutterstock.com/WildlifeWorld, Shutterstock.com/Albie Venter, Shutterstock.com/PHOTOCREO Michal Bednarek, Bernard DUPONT, CC BY-SA 2.0, via Wikimedia Commons; third row left to right: Shutterstock.com/SomprasongWittayanupakorn, Shutterstock.com/Ian Dyball, Shutterstock.com/Ondrej Prosicky, Shutterstock.com/Thomas Retterath; fourth row left to right: iStock.com/Ivan_Sabo, iStock.com/WLDavies, iStock.com/Iain Tall.